FLUID MECHANICS AND HYDRAULIC MACHINE

FLUID MECHANICS AND HYDRAULIC MACHINE

Manish Saxena

ANMOL PUBLICATIONS PVT. LTD.
NEW DELHI-110 002 (INDIA)

ANMOL PUBLICATIONS PVT. LTD.
Regd. Office: 4360/4, Ansari Road, Daryaganj,
New Delhi-110002 (India)
Tel.: 23278000, 23261597, 23286875, 23255577
Fax: 91-11-23280289
Email: anmolpub@gmail.com
Visit us at: www.anmolpublications.com

Branch Office: No. 1015, Ist Main Road, BSK IIIrd Stage
IIIrd Phase, IIIrd Block, Bangalore-560 085 (India)
Tel.: 080-41723429 • Fax: 080-26723604
Email: anmolpublicationsbangalore@gmail.com

Fluid Mechanics and Hydraulic Machine

First Edition, 2010

ISBN 978-81-261-4482-2

PRINTED IN INDIA

Printed at Mehra Offset Press, Delhi.

Contents

Preface

Fluid Mechanics and Hydraulic Machines is helpful text-cum-reference book for University students who are pursuing their Mechanical, Hydraulic and Electrical Engineering Courses from various Indian or Foreign Universities. The book explains the various properties of fluid mechanics, fluid dynamics, flow over notches, and pipe turbulent flow with centrifugal and reciprocating pumps.

The textbook basically concerned with the behaviour of liquids and gases and is important for many branches of engineering such as biomechanics, oceanography, chemical, aeronautical, mechanical, and civil engineering including environmental studies. The book examines various forms of energy, especially thermo and hydro power, and provides outlines of various types of power plants. An outstanding feature of the book is the classification of fluid mechanics.

Author

Chapter 1

Fluid and Continuum

FLUID MECHANICS

Continuum Mechanics is a means of studying the deformation and flow of matter by ignoring its particulate nature. It is a simplification that makes it possible to investigate the movement of matter on scales larger compared to typical distances between particles. Thus, in studying the movement of air in the atmosphere the fact that is it made up of atoms is ignored. It would, in any case, be impossible to predict the motion of each atom. Fortunately, when asking questions such as "How warm will it be tomorrow?" we are only interested in the average behaviour of large numbers of atoms, not in their individual motion. Similarly, in the study of mud flow the movement of industrial clay particles is ignored. At larger scales the evolution of galaxies can also be described using continuum models instead of trying to predict the movement of every planet, asteroid, and star.

This approach is used to study a huge range of phenomena, including the flow of air and water as well as in studies of rock slides (the individual particles being large rocks), snow avalanches, blood flow, and even galaxy evolution.

The Continuum and Fluid Mechanics courses in the Applied Mathematics department at the University of Waterloo. We hope that the material contained in these pages enriches the experience of those taking our courses (AM 361 and AM 463). At the same time, for those who are at an earlier point in their university education, we hope to show you both the richness of our field of study as well as illustrate how the courses you take earlier in your career tie in with what we do in our upper year courses and in our research.

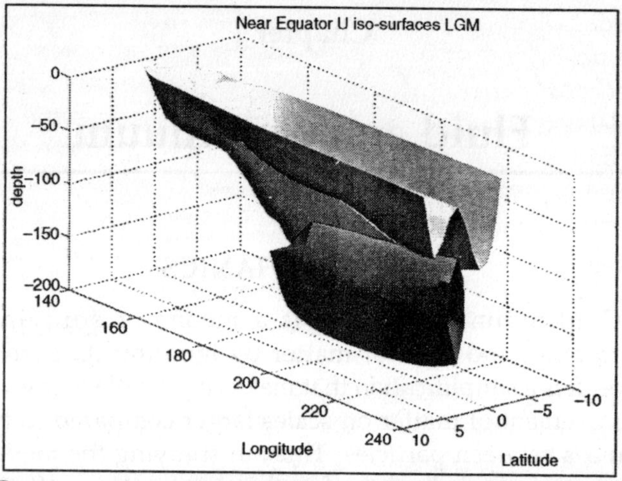

Continuum and Fluid Mechanics each have a long history. Indeed we take some past achievements nearly for granted whenever we fly in an airplane or turn on the television to check the weather report. Yet it is likely true that the wide availability of computers for research purposes makes the present a new Golden Age for the study of both fluid and continuum mechanics. It is perhaps easier to define fluid mechanics than continuum mechanics. After all we are surrounded, indeed bathed in, fluids for the entirety of our lives. This is because at the level of their gross, or macroscopic, behaviour (how they flow, for example) both gases and liquids can be considered to be fluids.

However, the actual mathematical description of fluid flow can be vastly different depending on the scales of interest. Thus, while the aerodynamics of a bumble bee's flight and the evolution of a hurricane are technically governed by the same set of equations, the simplifications made to reflect the dominant physics of each situation are wildly different.

As broad as fluid mechanics is, continuum mechanics can be considered to be even broader. This is because it is concerned with the general theory of equations that govern fluids, solids and all the exotic materials in between, from a general, and often abstract, point of view. Moreover, since much of the materials that surround us and make much of modern life possible (plastics being the most obvious example) are neither fluid nor solid, but something in between, there is also an intensely practical side to continuum mechanics. Indeed, the study of materials ranges across many disciplines in both the pure and applied sciences. One of the things we hope to convey throughout these web materials is the particular nature of studying materials within an applied mathematics setting.

VORTICES

Vortices are common features in fluid flows, They have important implications for the flight of airplanes where they are generated by boundary layer processes. Wing tip vortices are a limiting factor for the frequency with which airplanes can take off. An example of a much larger scale vortex is a hurricane. Flow at these scales is strongly affected by the earth's rotation: the flow around a hurricane is counterclockwise in the northern hemisphere and clockwise in the southern hemisphere.

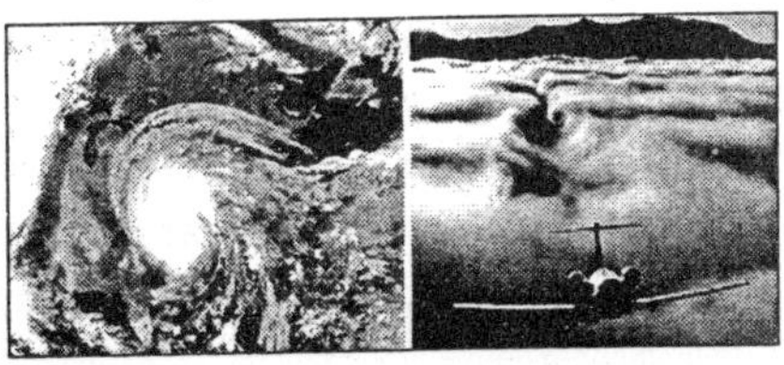

Fig. CessnaDownwash. Photo Credit: Gallery of Fluid Motion, Physics of Fluids A, Vol. 5, September 1993. Contribution by Hiroshi Higuchi (Syracuse University).

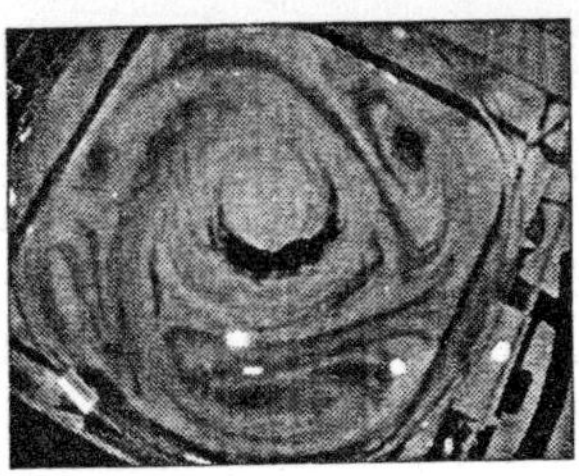

INTERNAL GRAVITY WAVES

Internal gravity waves are waves that occur in density stratified fluids in the presence of a gravitational field. They arise as gravitational restoring forces act on vertically displaced fluid. Internal gravity waves are all around us, but difficult to see.

They are ubiquitous in stratified lakes, in the ocean, and in the atmosphere. They are easily seen in a glass jar containing cooking oil and water. Internal waves are the waves you see on the interface. Better yet, try a layer of fresh water over salty water dyed with food colouring.

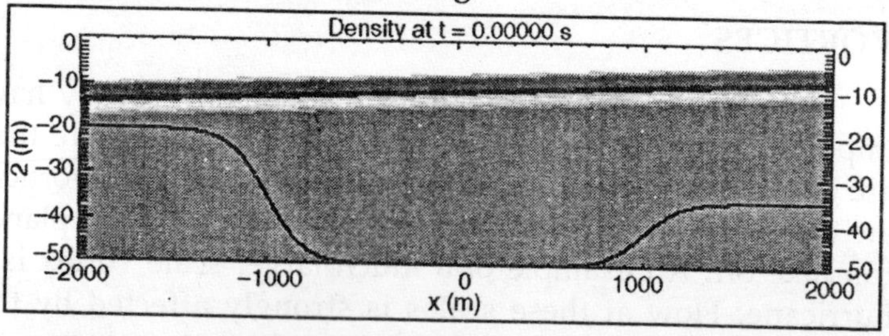

This animation shows how internal waves can be generated in a lake. Daytime winds drag warm surface water down wind forming a tilted epilimnion, or thermocline (the thin layer across which the water temperature changes). When the winds die down, the thermocline is released. The rocking basin scale standing wave, or seiche, evolves into a complex wave field, including trains of nonlinear solitary waves.

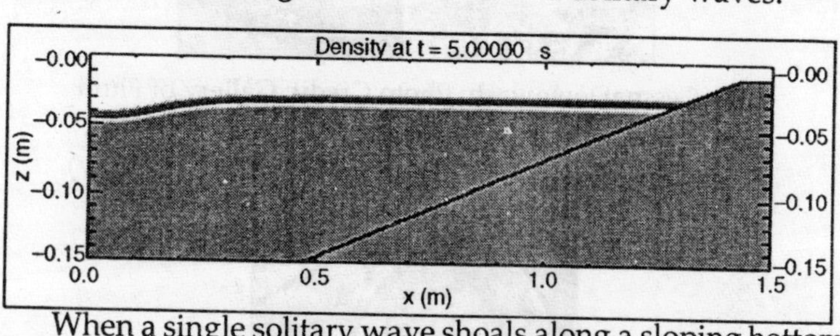

When a single solitary wave shoals along a sloping bottom it breaks, dissipating energy, mixing fluid and resuspending bottom sediments. This has important implications for biological and chemical processes.

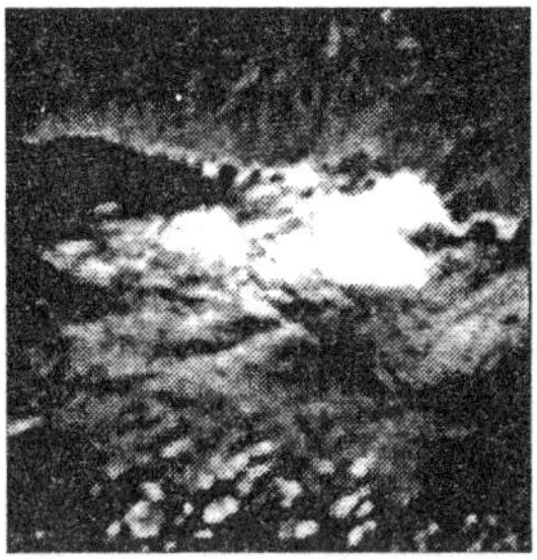

Internal Waves in the ocean and lakes lie beneath the surface. Their associated currents can modify the surface wave field, creating slicks on the surface which are visible from land or air. Here is a photo of internal wave slicks taken from the space shuttle.

Large amplitude internal waves in the ocean are usually generated by tidal flow over topographic features such as a bank edge.

CONCEPT OF FLUID MECHANICS

Fluid Mechanics is a sub discipline of Continuum Mechanics which focuses on the behaviour of fluids which include liquids and gases. Important examples for our everyday lives are water and air.

Beautiful photos of fluids in motion can be found in the Physics of Fluid Gallery. Some of our own images and discussions of fluids dynamics from a student perspective can be found here. Fluid mechanics is a branch of physics which concerns the study of fluids and the ways in which they interact with forces. Both liquids and gases are considered to be fluids for the purposes of this branch of science. Often, the field of fluid mechanics is divided into two more specific fields of study. These are fluid statics and fluid dynamics, which concern fluids at rest and fluids in motion, respectively. Fluid mechanics can involve highly complex mathematics, and the aid of modern computers has enhanced this science significantly.

The chronological roots of fluid mechanics go all the way back to at least the ancient Greeks. The Greek physicist and inventor Archimedes was the author of some of the first studies we know of which concern fluid statics, including the property of buoyancy. Persian philosophers in the Medieval time period coupled these ancient works with their own studies of fluid dynamics that acted as an early precursor to modern fluid dynamics. Such well-known historical figures as Leonardo da Vinci and Sir Isaac Newton, as well as others, made notable contributions to our understanding of fluid mechanics.

Every type of science starts out with basic, fundamental assumptions that govern the course of their study. Fluid mechanics is typically defined as having three basic premises or assumptions at its root. The first is the conservation of mass, which means that mass can neither be spontaneously created nor destroyed, although it may change forms. The second assumption, the conservation of momentum, is somewhat similar. This law states that the total momentum in a closed system is constant, and cannot spontaneously appear or disappear.

The third basic assumption governing fluid mechanics is what is known as the continuum hypothesis. This is a way of seeing fluids that does not take into account the presence of discrete molecules. Instead, a fluid's properties are assumed to vary in a continuous way from one point to the next.

Because it ignores the actual nature of small particles of matter, the continuum hypothesis is only an approximation used as a tool in calculations. It can result in a slightly inaccurate solution, but also in solutions that are very accurate under ideal circumstances. Other, more exact methods exist, but this hypothesis is often quite useful as a preliminary assumption. Many times, it can also be assumed that a given fluid is incompressible, meaning it cannot be compressed. This is only actually true of liquids, however, and not gases.

- Our courses Continuum Mechanics (AMATH 361) and Fluid Mechanics (AMATH 463) provide a unique opportunity for undergraduates to learn about this fascinating field. These courses combine mathematical analysis of the physical phenomena with computer visualization and demonstrations using equipment in our Fluids Lab

- The option in Earth Science, which can be combined with our AMATH programme, is designed for students who are interested in Fluid Mechanics and would like to learn how it is related to environmental issues.

Control Theory

- We have a fourth year course in Control Theory (AMATH 455), which builds on the foundational courses in ordinary differential equations and complex analysis. We also have a related course in the Calculus of Variations (AMATH 456).
- When taking the control theory course you will do some experiments in our Control Lab.
- If you find Control Theory interesting, consider enrolling in our programme Applied Mathematics with Engineering Electives. One of the areas of specialization is "Communication & Control". You could take a course in "Robot Dynamics & Control".

DYNAMICAL SYSTEMS

- Two of our foundation courses (AMATH 250 & AMATH 351) deal with methods for analyzing the solutions of differential equations. They provide the background for a fourth year course on dynamical systems.

Quantum Theory

- An introduction to the fascinating quantum world is provided by our courses AMATH 373/473 (Quantum Theory 1 and 2).
- AMATH 473, together with PHYS 467 (Introduction to Quantum Information Processing), will provide the foundation for those interested in research in Quantum Information Theory and Quantum Computing.
- If you find quantum theory interesting, consider enrolling in our programme in Mathematical Physics.

Scientific Computing

- Our courses AMATH 342 (Computational methods for DEs) And AMATH 442 (Computational methods for PDEs) give a solid introduction to the development of computational methods, their implementation in Matlab, and numerical simulations in a variety of areas of applications.
- If using computers to solve problems in science and technology appeals to you consider enrolling in our programme Scientific Computation/Applied Mathematics.

Have you ever wondered how airplanes fly (mathematicians 'proved' they could not around the turn of the 19th century!) or how weather predictions are made? Why dimples on a golf ball increase the distance it travels? Why curve balls curve? Do toilets really flush counterclockwise in the northern hemisphere and clockwise in the southern hemisphere? Why are surface water waves almost always nearly parallel to the beach? These are questions you can learn answers to if you study Applied Mathematics at Waterloo.

PHYSICAL PROPERTIES OF FLUIDS

INTRODUCTION

A treatment of fluid flow requires an understanding of the physical properties of a fluid which affect the motion, and an appreciation of the units in which these properties are measured.

The three most important fluid properties are the *density, viscosity* and *surface tension*. Density is a familiar concept and is treated briefly, although there is one subtlety that requires attention.

Viscosity is a concept that is generally understood intuitively, but some detailed consideration is necessary in order to establish a precise and useful definition. Each of these will be defined and viewed in terms of molecular concepts, and their dimensions will be examined in terms of mass, length and time (M, L and T).

DENSITY

The *density* p of a fluid is defined as its mass per unit volume and indicates its inertia or resistance to an accelerating force.

Thus,

$$\rho = \frac{\text{mass}}{\text{volume}} [=] \frac{M}{L^3}$$

in which the notation "[=]" is consistently used to indicate the *dimensions* of a quantity. It is usually understood in Eqn. 2.1 that the volume is chosen so that it is neither so small that it has no chance of containing a representative selection of molecules, nor is it so large that (in the case of gases) changes of pressure cause significant changes of density throughout the volume. A medium characterized by a density is called a *continuum,* and follows the classical laws of mechanics – including Newton's law of motion.

Densities of liquids

Density depends on the mass of an individual molecule and the number of such molecules that occupy a unit of volume. For liquids, density depends primarily on the particular liquid and, to a much smaller extent, on its temperature.

Representative densities are given in table. The accuracy of the values given in tables 2.1 through is adequate for the calculation needed for this course. However, if highly accurate values are needed, particularly at extreme conditions, then specialized information should be sought elsewhere.

Degrees A.P.I. (American Petroleum Institute) are related to specific gravity *s* by the formula:

$$^\circ\text{A.P.I.} = \frac{141.5}{s} - 131.5$$

Note that for water, °A.P.I. = 10, with correspondingly *higher* values for liquids that are *less* dense. Thus, for the crude oil listed in table, Eqn. indeed gives 141.5 / 0.851 – 131.5 = 35 °A.P.I.

Table. Specific Gravity, Densities, and Thermal Expansion Coefficients of Liquids at 20° C

Liquid	Sp. Gr.	Density, ρ		α
8	kg/m³	Ib_m/ft^3	°C⁻¹	
Acetone	0.792	792	49.4	0.00149
Benzene	0.879	879	54.9	0.00124
Crude oil, 35°API	0.851	851	53.1	0.00074
Ethanol	0.789	789	49.3	0.00112
Glycerol	1.26 (50°C)	1,260	78.7	—
Kerosene	0.819	819	51.1	0.00093
Mercury	13.55	13,550	845.9	0.0000182
Methanol	0.792	792	49.4	0.00120
n-Octane	0.703	703	43.9	—
n-Pentane	0.630	630	39.3	0.00161
Water	0.998	998	62.3	0.000207

DENSITY OF GASES

For ideal gases, $\rho V = nRT$, where p is the *absolute* pressure, V is the volume of the gas, n is the number of moles, R is the gas constant, and T is the *absolute* temperature.

If M_w is the molecular weight of the gas, it follows:

$$\rho = \frac{nM_w}{V} = \frac{M_w P}{RT}$$

Thus, the density of an ideal gas depends on the molecular weight, absolute pressure, and absolute temperature. Values of the gas constant R are given in table for various systems of units. Note the degrees Kelvin, formerly represented by "°K" is now more simply donated by "K".

Table. Volume of the Gas Constant, R

Value	Units
8.314	J/g-mole K
0.08314	liter bar/g-mole K
0.08206	liter atm/g-mole K
1.987	cal/g-mole K
10.73	psia ft³/lb-mole °R

0.7302	ft^3 atm/lb-mole °R
1,545	ft lb_f/lb-mole °R

For a *nonideal* gas, the compressibility factor Z (a function of *p* and *T*) is introduced into the denominator of Eqn. giving:

$$\rho = \frac{nM_W}{V} = \frac{M_W P}{ZRT}$$

Thus, the extent to which Z deviates from unity gives a measure of the nonideality of the gas.

The isothermal compressibility of a gas in defined as:

$$\beta = -\frac{1}{V}\left(\frac{\partial V}{\partial p}\right)_T$$

and equals – at constant temperature – the fractional *decrease* in volume caused by a unit *increase* in the pressure. For an ideal gas, $\beta = 1 / p$, the reciprocal of the absolute pressure.

The *coefficient of thermal expansion á* of a material is its isobaric fractional increase in volume per unit rise in temperature:

$$\alpha = \frac{1}{V}\left(\frac{\partial V}{\partial T}\right)_P$$

Since, for a given mass, density is inversely proportional to volume, it follows that for moderate temperature ranges (over which α is essentially constant) the density of most liquids is approximately a linear function of temperature:

$$\rho = \rho_0[1 - \alpha(T - T_0)]$$

where ρ_0 is the density at a reference temperature T_0. For an ideal gas, $\alpha = 1 / T$, the reciprocal of the absolute temperature.

The *specific gravity s* of a fluid is the ratio of the density ρ to the density ρ_{sc} of a reference fluid at some standard condition:

$$s = \frac{\rho}{\rho_{sc}}$$

For liquids, ρ_{sc} is usually the density of water at 4°C, which equals 1.000 g / ml or 1,000 kg / m^3. For *gases*, ρ_{sc} is sometimes taken as the density of air at 60°Fand 14.7 psia, which is approximately 0.0759 lb_m / ft^3, and sometimes at 0°C and one atmosphere absolute; since there is no *single* standard for gases, care must obviously be taken when interpreting published values. For *natural gas*, consisting primarily of methane and other hydrocarbons, the *gas gravity* is defined as the ratio of the molecular weight of the gas to that of air (28.8 lb_m / lb-mole). Values of the molecular weight M_w are listed in the table 2.3 for several commonly occurring gases, together with their densities at standard conditions of atmospheric pressure and 0°C.

Table. Gas Molecular Weights and Densities
The Latter at Atmospheric Pressure and 0°C

Gas	M_w	Standared kg/m³	Density lbm/ft³
Air	28.8	1.29	0.0802
Carbon dioxide	44.0	1.96	0.1225
Ethylene	28.0	1.25	0.0780
Hydrogen	2.0	0.089	0.0056
Methane	16.0	0.714	0.0446
Nitrogen	28.0	1.25	0.0780
Oxygen	32.0	1.43	0.0891

CONTINUUM HYPOTHESIS

The density provides a convenient means of illustrating the *continuum hypothesis*, which is one of the fundamental principles utilized in fluid mechanics. Density is defined as mass per unit volume, and it is commonly determined by measuring either the mass of a known volume or the volume displaced by a known mass. We recognize, however that density can vary with position in a process, and a more precise definition is to consider a series of small volumes Δυ, with corresponding small masses ΔM, and to define the density ρ by the limiting process.

$$\rho = \lim_{\Delta\upsilon \to 0} \frac{\Delta M}{\Delta \upsilon}$$

Alternatively, if the density ρ is a function of position, (x, y, z), then the mass contained in a specified volume is the integral of the density over that volume:

$$M = \int \rho(x,y,z)\,d\upsilon \equiv \iiint \rho(x,y,z)\,dx\,dy\,dz$$

Now, Eqns have implicitly used the continuum hypothesis, for they assume that the density is a concept that has meaning for any volume, no matter how small.

Yet we know that when the volume is of molecular dimensions we have very rapid spatial variations between dense and non-dense regions, depending on how close we are to a molecule.

Hence, the notion of a density function that can be defined by a limiting process such as that of Eqn. as above is questionable.

The continuum hypothesis assumes that mathematical limits for volumes tending to zero are reached over a scale that remains large compared to molecular dimensions.

Thus, variations on a molecular scale can be ignored, since they average out over all length scales of interest. Clearly, no result derived using the continuum hypothesis should be expected to apply a rarified gas, where the distance between molecules may be large.

VISCOSITY

The *viscosity* is a measure of resistance to flow. It we tip over a glass of water on the table, the water will spill out before we can stop it.

If we tip over a jar of honey, we probably can set it upright again before much honey flow out; this is because the honey has much greater resistance to flow, more viscosity, than water. A more precise definition of viscosity is possible in terms of following experiment:

Consider two long, solid plates separated by a thin film of fluid . If we slide the upper plate steadily in the x direction with velocity V_o, a force is needed to overcome the friction in the fluid between the plates. This force is different for different velocities, plate sizes, fluids, and distances between the plates. We can eliminate the effect of different plate sizes, however, by measuring the force per unit area of the plate, which we define as the shear stress τ.

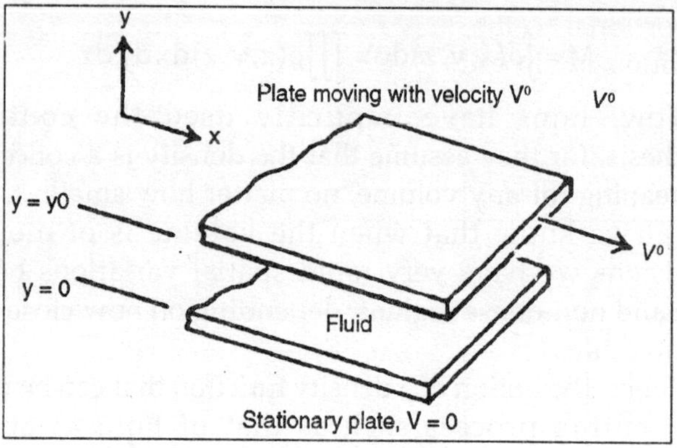

Fig. The Sliding Plate Experiment.

It has been demonstrated experimentally that at low values of V_o the velocity profile in the fluid between the plates is linear, i.e.,

$$V = \frac{V_o y}{y_o}$$

so that,

$$\frac{dV}{dy} = \frac{V_o}{y_o}$$

It has also been demonstrated experimentally that for most fluids the results of this experiment can be shown most conveniently on a plot τ versus dV / dy. As shown here, dV / dy is simply is velocity divided by a distance. In more complex geometries, it is the limiting value of such a ratio at a point. It is commonly called the *shear rate, rate of strain,* and *rate of shear deformation,* which all means exactly the same thing. Four different kinds of curve are shown as experiment results in the figure.

All four are observed in nature. The behaviour most common in nature is that represented by the straight line through the origin. This line is called *Newtonian* because it is described by Newton's law of viscosity,

$$\iota = \upsilon \frac{dV}{dy}$$

This equation says that the shear stress τ is linearly

proportional to the velocity gradient dV / dy. It is also the definition of viscosity, because we can rearrange it as,

$$\upsilon \frac{\iota}{dV/dy}$$

Here μ is called the *viscosity* or the coefficient of *viscosity*.

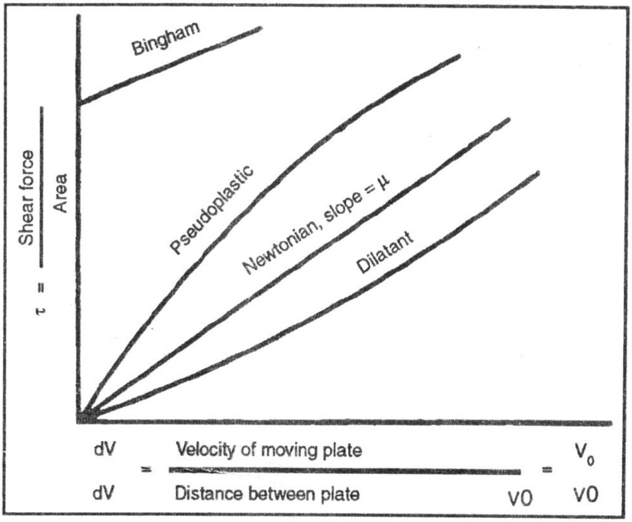

Fig. Possible outcomes of the Sliding-plate Experiment at Constant Temperature and Pressure.

For fluids such as air, the value of μ is very low; therefore, their observed behaviour is represented in Fig above by a straight line through the origin, very close to the dV/dy axis. For fluid such as corn syrup, the value of μ is very large, and the straight line through the origin is close to the τ axis.

Fluids that exhibit this behaviour in the sliding plate experiment (i.e., fluids that obey Newton's law of viscosity) are called *Newtonian fluids.*

All others are called *nonnewtonian fluids.* Which fluids are Newtonian? All gases are Newtonian. All liquids for which we can write a simple chemical formula are Newtonian, such as water, benzene, ethyl alcohol, carbon tetrachloride, and hexane. Most solutions of simple molecules are Newtonian, such as aqueous solutions of inorganic salts and of sugar. Which fluids are nonnewtonian? Generally, nonnewtonian fluids are complex mixtures: slurries, pastes, gels, polymer solutions, etc. Most nonnewtonian fluids are composed of molecules or particles that are much larger than water molecules, such as the sand grains in a mud or the collagen molecules in gelatin, which are thousands or millions of times larger than water molecules.

In discussing non newtonian fluids, we must agree on what we mean by viscosity. If we retain the definition given by Eqn. then the viscosity can no longer be considered a constant independent of *dV/dy* for a given temperature, but must be considered a function of *dV/dy*. This is shown in figure below.

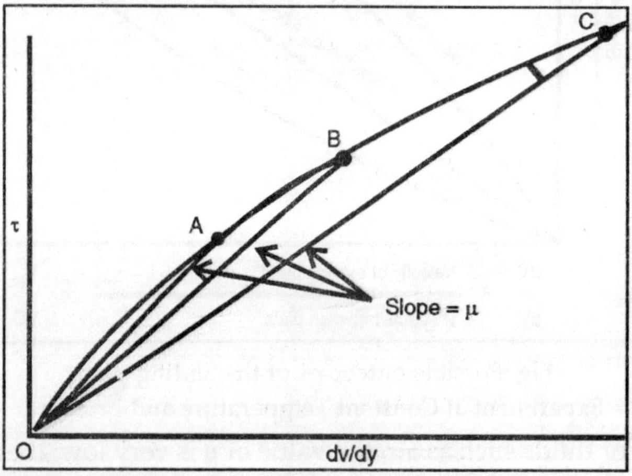

Fig. Viscosity of Nonnewtonian Fluids at Constant Temperature

Here each of lines OA, OB, and OC have slope μ, so the viscosity is sdecreasing with increasing *dV/dy*.

Using this definition, we can observe that there are three common types of nonnewtonian fluid:

- *Bingham fluids*: Sometimes called *bingham plastics*, resist a small shear stress indefinitely but flow easily under larger shear stresses. One may say that at low stresses the viscosity is infinite and that at higher stresses the viscosity decreases with increasing velocity gradient. Examples are bread dough, tooth paste, jellies, and some slurries.

- *Pseudoplastic fluids:* Show a viscosity that decreases with increasing velocity gradient. Examples are most slurries, muds, polymer solutions, solutions of natural gums, and blood.
- *Dilatant fluids:* Show a viscosity that increases with increasing velocity gradient. They are uncommon, but starch suspensions behave in this way.

The basic unit of viscosity is the poise, where 1P = 1g / (cm.s) = 0.1 Pa.s = 6.72×10^{-2} lbm / (ft.s). It is widely used for materials such as high-polymer solutions and molten polymers.

However, it is too large a unit for most common fluids. By sheer coincidence the viscosity of pure water at about 68°F is 0.01 P; for that reason the common unit of viscosity in the United States is the centipoises, where

$$1\text{ cP} = 0.01\text{ P} = 0.01\text{ g}/(\text{cm.s}) = 6.72 \times 10^{-4}\text{lb}_m/(\text{ft.s}) = 0.001\text{Pa.s.}$$

Hence, the viscosity of a fluid, expressed in centipoises, is the same as the ratio of its viscosity to that of water at room temperature.

Kinematic Viscosity

In many engineering problems, viscosity appears only in the relation of viscosity divided by density. Therefore, to save writing, we define,

$$\text{Kinematic Viscosity} = \upsilon = \frac{\mu}{\rho}$$

The most common unit of kinematic viscosity is the centistokes (cSt):

$$1\text{cSt} = \frac{1\text{cP}}{1\text{g}/\text{cm}^3} = 1.08 \times 10^{-5}\frac{\text{ft}^2}{\text{s}} = 10^{-6}\frac{\text{m}^2}{\text{s}}$$

(The kinematic viscosity of water at room temperature is 1cSt.) To avoid confusion over which viscosity is being used, some writers refer to the viscosity μ as the absolute viscosity.

SURFACE TENSION

Surface tension is the tendency of the surface of a liquid

to behave like a stretched elastic membrane. There is a natural tendency for liquids to minimize their surface area. The obvious case is that of a liquid droplet on a horizontal surface that is not wetted by the liquid – mercury on glass, or water on a surface that also has a thin oil film on it.

For small droplets, such as those on the left of the Fig. below the droplet adopts a shape that is almost perfectly spherical, because in this configuration there is the least surface area for a given volume.

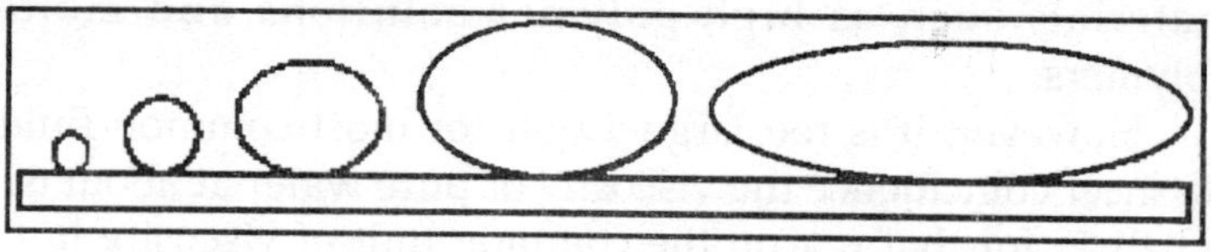

Fig. The Larger Droplets are Flatter Because Gravity is Becoming more Important than Surface Tension.

For larger droplets, the shape becomes somewhat flatter because of the increasingly important gravitational effect, which is roughly proportional to a^3, where a is the approximate droplet radius, whereas the surface area is proportional only to a^2.

Thus, the ratio of gravitational to surface tension effects depends roughly on the value of $a^3/a^2 = a$, and is therefore increasingly important for the larger droplets, as shown to the right in the Figure above.

Overall, the situation is very similar to that of a water-filled balloon, in which the water accounts for the gravitational effect and the balloon acts like the surface tension.

A fundamental property is the *surface energy,* which is defined with reference to the Figure a. A molecule I, situated in the *interior* of the liquid, is attracted equally in all directions by its neighbors.

However, a molecule S, situated in the *surface,* experiences a net attractive force into the bulk of the liquid. (The molecule S.) Therefore, work has to be done against such a force in bringing an interior molecule to the surface. Hence, an energy σ, called the *surface energy,* can be attributed to a unit area of the surface.

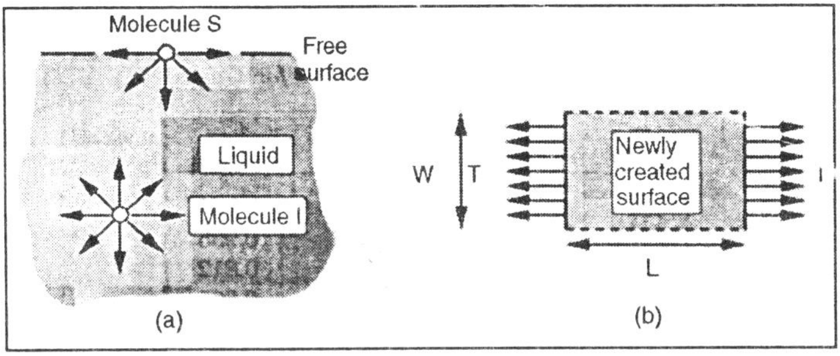

Fig. (a) Molecules in the Interior and Surface of a Liquid; (b) Newly Created Surface Caused by Moving the Tension T Through a Distance L.

An equivalent viewpoint is to consider the surface tension *T* existing per unit distance of a line drawn in the surface, as shown in the Figure b. Suppose that such a tension has moved a distance *L*, thereby creating an area *WL* of fresh surface. The work done is the product of the force, *TW*, and the distance *L* through which it moves, namely *TWL*, and this must equal the newly acquired surface energy σWL Therefore, $T = \sigma$; both quantities have units of force per unit distance, such as N/m, which is equivalent to energy per unit area, such as J/m^2.

UNITS AND SYSTEMS OF UNITS

Mass, weight, and force

The *mass M* of an object is a measure of the amount of matter it contains and will be constant, since it depends on the number of constituent molecules and their masses. On the other hand, the *weight w* of the object is the gravitational force on it, and is equal to *Mg*, where *g* is the local gravitational acceleration. Mostly, we shall be discussing phenomena occurring at the surface of the earth, where g is approximately 32.174 ft/s^2 = 9.807 m/s^2= 980.7 cm/s^2 These values are simply taken as 32.2, 9.81 and 981, respectively.

Table. (Representative Units of Force)

System	Units of Force	Customary Name
SI	kg m/s^2	Newton

CGS	g cm/s^2	Dyne
FPS	lbm ft/s^2	Poundal

Newton's second law of motion states that a force *F* applied to a mass *M* will give it an acceleration *a*:

F = ma

from which is apparent that force has dimensions ML/T^2. Table 2.4 gives the corresponding *units* of force in the SI (meter / kilogram / second), CGS (centimeter / gram / second), and FPS (foot / pound / second) systems. The poundal is now an archaic unit, hardly ever used. Instead, the *pound force*, ib$_f$, is much more common in the English system; it is defined as the gravitational force on 1 ib$_m$, which, if left to fall freely, will do so with an acceleration of 32.2 ft/s^2.

Hence:

$$1\,\text{lb}_f = 32.2\,\text{lb}_m \frac{\text{ft}}{\text{s}^2} = \text{poundals}$$

When using lb$_f$ in the ft, lb$_m$, s (FPS) system, the following conversion factor, commonly called "g_c", will almost invariably be needed:

$$g_c = 32.2 \frac{\text{lb}_m \text{ft}}{\text{lb}_f \text{s}^2}$$

Some writers incorporate g_c into their equations, but this approach may be confusing since it virtually implies that one particular set of units is being use, and hence tends to rob the equations of their generality. Why not, for example, also incorporate the conversion factor of 144 in^2/ft^2 into equations where pressure is expressed in lb$_f$/in^2? We prefer to omit all conversion factors in equations, and introduce them only as needed in evaluating expressions numerically. If the reader is in any doubt, units should *always* be checked when performing calculations.

SI Units: The most systematically developed and universally accepted set of units occurs in the *SI* units or *Systeme International d'Units*; the subset we mainly need is shown in table

The basic units are again the meter, kilogram, and second (m, kg, and s); from these, certain *derived* units can also be obtained. Force (kg m / s^2) has already been discussed; energy is the product of force and length; power amounts to energy per unit time; surface tension is energy per unit area or force per unit length, and so on.

Some of the units have names, and these, together with their abbreviations, are also given in table.

Table. (SI Units)

Psysical Quantity	Name of Unit	Symbol for Unit	Definition of Unit
Basic Units			
Length	meter	m	–
Mass	kilogram	kg	–
Time	second	s	–
Temperature	degree Kelvin	K	–
Supplementary unit			
Plane angle	radian	rad	–
Derived Units			
Acceleration			m/s^2
Angular Velocity			rad/s
Density			kg/m^3
Energy	joule	J	kg/m^3
Force	newton	N	$kg\ m^2/s^2$
Kinematic viscosity			
Power	watt	W	m^2/s^2
Pressure	pascal	Pa	$kg/m^2/s^3$ (J/s)
Velocity			m/s
Viscosity			kg/m s

Tradition dies hard, and certain other "metric" units are so well established that they may be used as *auxiliary* units; these are shown in table.

The *gram* is the classic example. Note that the basic SI unit of mass (kg) is even represented in terms of the gram, and has not yet been given a name of its own!

Table. (Auxiliary Units Allowed in Conjunction with SI Units)

Physical Quantity	Name of Unit	Symbol for Unit	Definition of Unit
Area	hectare	ha	10^4 m^2
inematic viscosity	stokes	St	10^{-4} m^2/s
Length	micron	μm	10^{-6} m
Mass	tonne	t	10^3 kg = Mg
	gram	g	10^{-3} kg = g
Pressure	bar	bar	10^5 N/m^2
Viscosity	poise	P	10^{-1} kg/ m s
Volume	liter	l	10^{-3} m^3

Table shows some of the acceptable prefixes that can be used for accommodating both small and large quantities. For example, to avoid an excessive number of decimal places, 0.000001 *s* is normally better expressed as 1 μs (one microsecond). Note also, for example, that 1 μkg should be written as 1 mg-one prefix being better than two.

Table. (Prefixes for Fractions and Multiples)

Factor	Name	Symbol	Factor	Name	Symbo
10^{-12}	pico	p	10^3	kilo	k
10^{-9}	nano	n	10^6	mega	M
10^{-6}	micro	μ	10^9	giga	G
10^{-3}	milli	m	10^{12}	tera	T

Some of the more frequently used conversion factors are given in the table

Example: Units Conversion

Part 1: Express 65 mph in (a) ft/s, and (b) m/s

Solution:

(a) $$65\frac{\text{mile}}{\text{hr}}\times\frac{1}{3.600}\frac{\text{hr}}{\text{s}}\times 5{,}280\frac{\text{ft}}{\text{mile}}=95.33\frac{\text{ft}}{\text{s}}$$

(b) $$95.33\frac{\text{ft}}{\text{s}}\times 0.3048\frac{\text{m}}{\text{ft}}=29.06\frac{\text{m}}{\text{s}}$$

Part 2: The density of 35°API crude oil is $53.11b_m/ft^3$ at

68°F and its viscosity is 32.8 lb_m/ft hr. What are its density, viscosity, and kinematic viscosity in SI units?

Solution:

$$\rho = 53.1 \frac{lb_m}{ft^s} \times 0.4536 \frac{kg}{lb_m} \times \frac{1}{0.3048^2} \frac{ft^s}{m^s} = 851 \frac{kg}{m^s}$$

$$\mu = 32.8 \frac{lb_m}{ft\,hr} \times \frac{1}{2.419} \frac{centipoises}{lb_m / ft\,hr} \times 0.01 \frac{poises}{centipoises} = 0.361\,poises$$

Or, converting to SI units, noting that P is the symbol for poise, and evaluating v:

$$m = 0.361\ P \times 0.1 \frac{kg/ms}{P} = 0.0136 \frac{kg}{ms}$$

$$\nu = \frac{\mu}{\rho} = \frac{0.0136\,kg/ms}{851\,kg/m^s} = 1.60 \times 10^{-5} \frac{m^2}{s} (=0.160\,st)$$

Example: Mass of Air in a Room

Estimate the mass of air in your classroom, which is 80 ft wide, 40 ft deep, and 12 ft high. The gas constant is R = 10.73 psia ft^3/lb – mole°R.

Solution: The volume of the classroom is:

$$V = 80 \times 40 \times 12 = 3.84 \times 10^4\ ft^3$$

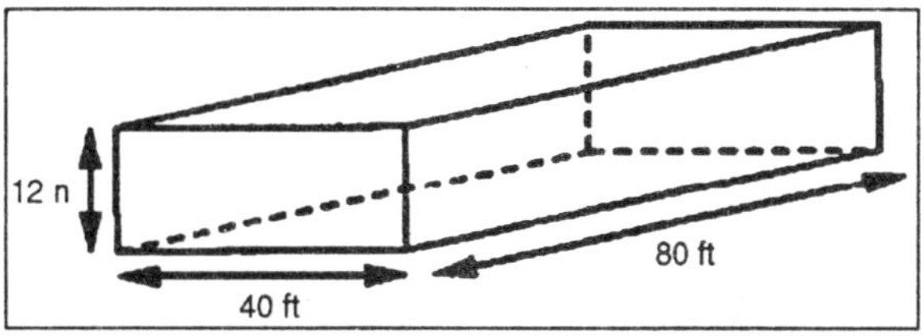

Fig. Assumed Dimensions of Classroom

If the air is approximately 20% oxygen and 80% nitrogen, its mean molecular weight is $M_W = 0.8 \times 28 + 0.2 \times 28.8$ lb_m/lb – Mole. From the gas law, assuming an absolute pressure of p = 14.7 psia and a temperature of 70 °F = 530 °R.

DENSITY OF LIQUIDS

FROM FLOW MEASUREMENT

The density of a liquid can vary with temperature and

pressure. In nearly all cases the effect of pressure is so small that it can be ignored completely. Temperature, however, does have a significant effect on density.

In general, liquids expand as temperature increases, and, therefore, the density decreases. Much data is available to give the density of various liquids at different temperatures. Sometimes it is in tabular form, as in Table .

This makes it very easy to readthevalues. In other cases there is only a listing of equations for the volumetric expansion of the liquid. To use this data, the density must be known at one temperature. Then, by applying the volumetric expansion coefficient values in proper ratio,the density at another temperature can be calculated.

Often graphical density data is available. Figure 3-3 is an example. These graphs are sometimes difficult to read accurately but can often serve as useful information. Figure 3-3 shows densities for a variety of liquids. Notice that over large portions of the graph the lines are quite straight. Therefore, over a designated range of temperature, it is quite possible to write a simple equation that expresses density as a linear function of temperature.

For example, the equation for 65°Brix (Bx) sucrose solution over the range of 20°C to 60°C is:

$$\text{Density in kg/l} = 1.33 - 0.058T_{°C} \text{ -1)}$$

The Brix scale is a hydrometer scale for sugar solutions in which its readings at a specified temperature represent percentages by weight of sugar in the solution.

Table.Example of Tabular Density Data (Density Measured by gm/cc)

Barium Chloride Solution in Water ($BaCl_2$)						
%BaCI2	**0°C**	**20°C**	**40°C**	**60°C**	**80°C**	**100°C**
2	1.0181	1.0159	1.0096	1.0004	0.9890	0.9755
4	1.0368	1.0341	1.0275	1.0181	1.0066	0.9931
8	1.0760	1.0721	1.0648	1.0551	1.0434	1.0299
12	1.1178	1.1128	1.1047	1.0948	1.0827	1.0692
16	1.1627	1.1564	1.1478	1.1373	1.1249	1.1113
20	1.2105	1.2031	1.1938	1.1828	1.1720	1.1563

24	———	1.2531	1.2430	1.2316	1.2186	1.2045
26	———	1.2793	1.2688	1.2571	1.2440	1.2798

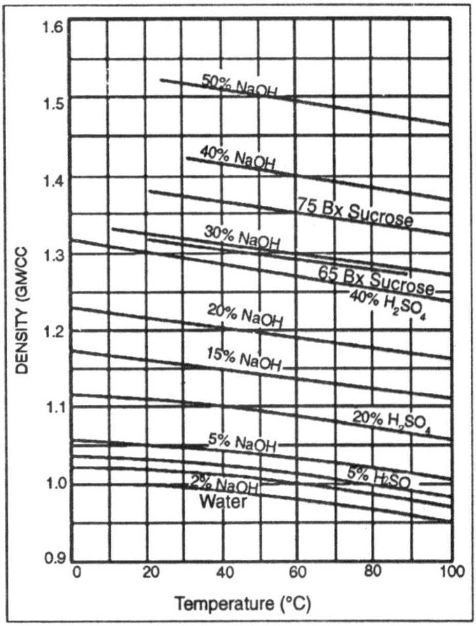

FLUID VISCOSITY

FROM FLOW MEASUREMENT

Viscosity is the internal resistance of a fluid to flow. Water has low viscosity, while syrup has high viscosity. When syrup is cold it is very much more viscousthan when it is hot. Viscosity is highly temperature dependent.

ABSOLUTE AND KINEMATIC VISCOSITY

There are two ways to measure the viscosity of a fluid. One is to rotate a cup ordisk in the fluid and measure the rotational speed along with the torque required tokeep it rotating.

The density of the fluid doesn't play any part in the measurement.Viscosity is the torque divided by the speed and divided bythe contacting surfacearea. When measured this way, the viscosity is called "absolute viscosity." Itsunits are centipoise, pascal-seconds, or lb/ft-sec.

Another way to measure viscosity is to allow a certain volume of fluid to drain by gravity out of a container through a capillary tube or some other restriction.The time to drain is directly related to viscosity and is often reported in "seconds."

CONTINUITY EQUATION

Even though a detailed analysis of fluid flow can be extremely difficult, the basic concepts involved in fluid flow problems are fairly straightforward. These basic concepts can be applied in solving fluid flow problems through the use of simplifying assumptions and average values, where appropriate.

Even though this type of analysis would not be sufficient in the engineering design of systems, it is very useful in understanding the operation of systems and predicting the approximate response of fluid systems to changes in operating parameters. The basic principles of fluid flow include three concepts or principles; the first two of which the student has been exposed to in previous manuals. The first is the principle of momentum(leading to equations of fluid forces) which was covered in the manual on Classical Physics.

The second is the conservation of energy (leading to the First Law of Thermodynamics) which was studied in thermodynamics. The third is the conservation of mass (leading to the continuity equation) which will be explained in this module.

PROPERTIES OF FLUIDS

A *fluid* is any substance which flows because its particles are not rigidly attached to one another. This includes liquids, gases and even some materials which are normally considered solids, such as glass. Essentially, fluids are materials which have no repeating crystalline structure. Several properties of fluids were discussed in the Thermodynamics section of this text.

These included temperature, pressure, mass, specific volume and density. *Temperature* was defined as the relative measure of how hot or cold a material is. It can be used to predict the direction that heat will be transferred. *Pressure* was defined as the force per unit area. Common units for pressure are pounds force per square inch (psi).

Mass was defined as the quantity of matter contained in a body and is to be distinguished from weight, which is measured by the pull of gravity on a body. The *specific volume* of a substance is the volume per unit mass of the substance. Typical units are ft^3/lbm. *Density*, on the other hand, is the mass of a substance per unit volume.

Typical units are lbm/ft^3. Density and specific volume are the inverse of one another. Both density and specific volume are dependant on the temperature and somewhat on the pressure of the fluid. As the temperature of the fluid increases, the density decreases and the specific volume increases.

Since liquids are considered incompressible, an increase in pressure will result in no change in density or specific volume of the liquid. In actuality, liquids can be slightly compressed at high pressures, resulting in a slight increase in density and a slight decrease in specific volume of the liquid.

The properties outlines below are general properties of fluids which are of interest in engineering. The symbol usually used to represent the property is specified together with some typical values in SI units for common fluids. Values under specific conditions (temperature, pressure etc.) can be readily found in many reference books. The dimensions of each unit is also give in the MLT system (see later in the section on dimensional analysis for more details about dimensions.)

Density

The density of a substance is the quantity of matter contained in a unit volume of the substance. It can be expressed in three different ways.

Mass Density

Mass Density, ρ, is defined as the mass of substance per unit volume.

Units: Kilograms per cubic metre, kg/m^3 (or kgm^{-3})

Dimensions: ML^{-3}

Typical values: Water = 1000$^{kgm^{-3}}$, Mercury = 13546$^{kgm^{-3}}$ Air = 1.23$^{kgm^{-3}}$, Paraffin Oil = 800$^{kgm^{-3}}$. (at pressure =1.013 = 1.013 × 10^{-5} Nm^{-2} and Temperature = 288.15 K.)

Specific Weight

Specific Weight ω, (sometimes, and sometimes known as *specific gravity*) is defined as the weight per unit volume.

or

The force exerted by gravity, g, upon a unit volume of the substance. The Relationship between g and ω can be determined by Newton's 2[nd]Law, since weight per unit volume = mass per unit volume g,

$$\omega = \rho g$$

Units: Newton's per cubic metre, N/M^3 (or Nm^{-3})

Dimensions: $ML^{-2}T^{-2}$.

Typical values:

Water = 9814 Nm^{-3}, Mercury = 132943 Nm^{-3}, Air = 12.07 Nm^{-3}, Paraffin Oil =7851 Nm^{-3}

Relative Density

Relative Density, σ, is defined as the ratio of mass density of a substance to some standard mass density. For solids and liquids this standard mass density is the maximum mass density for water (which occurs at 4°c) at atmospheric pressure.

$$s = \frac{\rho_{\text{sibstamce}}}{\rho_{H_2O(\omega 4°c)}}$$

Units: None, since a ratio is a pure number.

Dimensions: 1.

Typical values: Water = 1, Mercury = 13.5, Paraffin Oil =0.8.

Viscosity

Viscosity, m, is the property of a fluid, due to cohesion and interaction between molecules, which offers resistance to sheer deformation. Different fluids deform at different rates under the same shear stress. Fluid with a high viscosity such as syrup, deforms more slowly than fluid with a low viscosity such as water. All fluids are viscous, "Newtonian Fluids" obey the linear relationship given by Newton's law of viscosity which we saw earlier,

$$\tau=\mu\frac{du}{dy},$$

where τ is the shear stress,

Units: Nm^{-2}; $kg\ m^{-1}s^{-2}$

Dimensions $ML^{-1}T^{-2}$.

$$\frac{du}{dy}$$

is the velocity gradient or rate of shear strain, and has

Units: radianss^{-1}

Dimensions t^{-1}

μ is the "coefficient of dynamic viscosity"-see below.

Coefficient of Dynamic Viscosity

The Coefficient of Dynamic Viscosity, μ, is defined as the shear force, per unit area, (or shear stress τ), required to drag one layer of fluid with unit velocity past another layer a unit distance away.

$$\mu=\tau\frac{du}{dy}=\frac{Force}{Area}\frac{Velocity}{Distance}=\frac{Force\times Time}{Area}=\frac{Mass}{Length\times Area}$$

Units: Newton seconds per square metre, Nsm^{-2} or Kilograms per meter per second, $kgm^{-1}\ s^{-1}$. (Although note that μ is often expressed in Poise, P, where 10 P = 1 $kgm^{-1}\ s^{-1}$.)

Typical values: Water =1.14×10^{-3} Air =$1.78\times10^{-5}\ kgm^{-1}s^{-1}$, Mercury = 1.552 $kgm^{-1}\ s^{-1}$,

Paraffin Oil =1.9 $kgm^{-1}\ s^{-1}$.

Kinematic Viscosity

Kinematic Viscosity, ν, is defined as the ratio of dynamic viscosity to mass density.

$$\nu\frac{\mu}{\rho}$$

Units: square metres per second, m^2s^{-1} (Although note that n is often expressed in Stokes, St, where,

10^4 St = 1 m^2s^{-1}), St = 1 m^2s^{-1})

Dimensions: $L^2 T^{-1}$.

Typical values:

Water =1.14×10^{-6} m^{2}s^{-1}, Air = 1.46×10^{-5} m^{2}s^{-1}, Mercury =1.145×10^{-4} m^{2}s^{-1},

Paraffin Oil = 2.375×10^{-3} m^{2}s^{-1}

BUOYANCY

Buoyancy is defined as the tendency of a body to float or rise when submerged in a fluid. Weall have had numerous opportunities of observing the buoyant effects of a liquid. When we goswimming, our bodies are held up almost entirely by the water. Wood, ice, and cork float onwater. When we lift a rock from a stream bed, it suddenly seems heavier on emerging from thewater. Boats rely on this buoyant force to stay afloat. The amount of this buoyant effect wasfirst computed and stated by the Greek philosopher Archimedes. When a body is placed in afluid, it is buoyed up by a force equal to the weight of the water that it displaces.

RHEOLOGY OF FLUIDS

For most fluids, viscosity is not a constant but varies with the shear rate. Such fluids are called rate dependent. To check for rate dependent effects, a fluid is tested for shear stress or viscosity at a number of shear rates on the same viscometer. From this data the rheological constants from the Bingham and Power Law Models can be obtained or when this data is plotted the result is called a viscosity profile. A few examples are shown below.

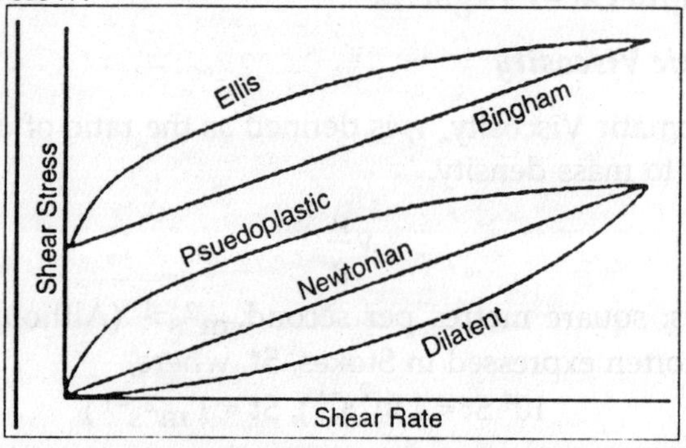

In a few systems, such as in water and oils, the shear stress and shear rate are in direct proportion. Such fluids have a constant viscosity. They are termed Newtonian after Newton, who first postulated their existence.

Most polymers are pseudoplastic. In this case, increased shear rate causes a progressive decrease in viscosity and inversely a decrease in shear rate will result in an increase in viscosity due to alignment of the long polymer chains along the flow lines. A few rare concentrated suspensions have been found to be dilatent.

These fluids increase in viscosity with increasing shear. Some fluids have a certain critical shear stress that must be exceeded before flow can begin. This critical shear stress is called the "yield value". If the fluid has essentially Newtonian flow after the yield value is exceeded, it is called a Bingham Plastic fluid. This model has been used extensively for circulating fluids in the oil industry.

Another model that has gained acceptance for pseudoplastic fluids is the Ostwald-deWaele-Nutting or Power Law model. If the fluid has a yield value and also follows the pseudoplastic model, it is referred to as an Ellis fluid and can be described by a simplified Hershel-Bulkley model. The Power Law and Hershel-Bulkley model have often been used to describe polymer solutions.

Besides the rate dependent effects discussed above, fluids can also exhibit time dependent effects. In these cases, the viscosity of the fluid tends to drop with continued shear at a constant rate. The following curve illustrates the two possible time dependent effects.

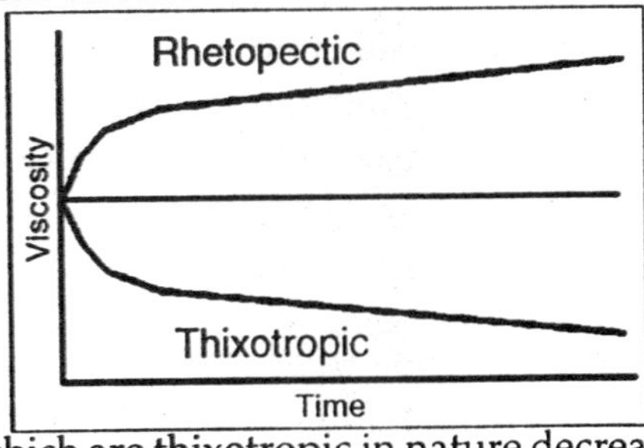

Fluids which are thixotropic in nature decrease in viscosity with time. Some polymers are thixotropic to a certain degree. In certain rare situations, highly concentrated suspensions can display rheopectic flow.

In these unusual cases, the viscosity increases with continued shear at a given rate. These models of rate and time dependent effects are designed to provide assistance in understanding fluid flow. Different models may suite the same fluid under various conditions. There are always a variety of possible explanations for any given situation.

Numerous viscometers are available for measuring the properties of a fluid. If it is desired to measure the viscosity of a sample and obtain data in basic rheological parameters (shear stress and shear rate) a viscometer that operates using one of three basic flow types; cone and plate, Colette, or capillary, must be used.

Empirical viscometers, not based on these flow types are widely used but the data from these can not be reduced to fundamental rheological measurements. Some of the more common types of viscometers are listed in Application of these principles to real fluid systems is not difficult. one must have an understanding of the critical property (or properties) of a system.

For example, a suspension that is designed to remain homogeneous must have sufficient viscosity at low shear rates in order to prevent solids settling. On the other hand, some applications require that the viscosity at high shear rates be as low as possible to reduce friction pressure losses. It is recognized that measurements at room temperature and pressure do not truly represent the fluid under down hole conditions, but extensive field studies have resulted in a high degree of success in predicting a fluids performance from this data. g system. It provides a useful addition to the "normal" steady shear measurements.

VISCO ELASTICITY

The distinction between solid and liquid materials is not as clear as the preceding would indicate. Most materials have some solid like character as well as some viscous properties.

It is easy to visualize the appearance differences between viscous liquids and gelled or cross-linked solids.

These fluids are characterized by taking the viscosity of liquids and the gel strength of solids. Many different viscometers and gel testers are sold, depending on the industry.

A problem arises when one tries to characterize a fluid that behaves like both a viscous liquid and a gel. Such fluids are called viscoelastic fluids. They can be pumped easily (although they have the appearance of a viscous liquid), but they are able to suspend small solid particles indefinitely (so have the characteristic of an elastic solid). In fact, molten polymers are viscoelastic to different degrees.

A way of quantifying the viscoelastic properties of materials is by measurement of their elastic modulus and viscous modulus. It is convenient to think of these as vector quantities located 90 degrees apart. Vector quantities have a magnitude (which is indicated by the length of the vector) as well as a direction. When the rheology is measured with a viscometer in steady shear, only the resultant vector of the viscous and elastic modulus is measured, this is usually called the shear stress.

Shear stress divided by the shear rate is called complex viscosity. Both of these fluids would have the same steady shear or complex viscosity since the resultant vector of the two moduli is the same magnitude.

However, viscoelastic measurement reveals that they are quite different in their nature. Much more the "viscosity" of the elastic fluid is due to its elastic modulus while the viscous fluid derives most of its "viscosity" from its viscous modulus. The elastic fluid is much better at suspending particles or stabilizing a foam.

The viscous fluid will slow down the rate of settling but the primary effect is an increase in friction pressure. If the solution has a significant elastic component, it will act like a solid and may actually be able to prevent settling.

These differences can only be revealed through viscoelastic measurements and performance comparisons.

There are laboratory instruments called rheometers that can simultaneously measure the viscous and gel nature of a fluid.

The rheometer used in this report is made by Rheometrics. It has a cup and bob configuration much like the Fann. Because the rheometer operates in an oscillating mode by applying a sinusoidally varying strain or deformation to the sample, it does not generate shear rate data in units of sec-1. The technically correct designation for its shear rate measurements called "frequency" are in terms of radians per second. Fortunately frequency in radians/second closely correlates with shear rate in sec-1 from the rotational viscometer.

The rheometer measures the resistance to the oscillation, and the phase delay of the resistance. A perfect elastic gel would have no phase delay, while a perfect viscous liquid would have a 90 degree phase delay.

All materials, have a phase angle somewhere between 0 and 90 degrees. The computer attached to the rheometer can qualify both the viscous nature, called the viscous modulus, and the gel nature, called the elastic modulus or elasticity.

The instrument also determines the vector sum of these two moduli which is the complex viscosity. While it is undoubtedly a sophisticated technique, viscoelastic measurements can add a great deal of understanding of today's polymers.

RHEOLOGY AND FLUID DYNAMICS

Rheology is the study of the way in which materials deform and flow. The materials may range from elastic solids to viscous liquids but in rheology interest is usually focussed on those materials that possess both elastic and viscous properties.

At the University of Plymouth, research into the determination of the rheological properties of materials, particularly liquids, has been carried out for many years, and was led by Eirian Jones (deceased), together with Mansel Davies, Jason Hughes, David Graham and Phil James.

In the 2001 RAE two of these staff joined the PRISM research group in the School of Engineering, and formed part of a 4 rated submission to the Mechanical, Aeronautical and Manufacturing Engineering Unit of Assessment (Unit 30).

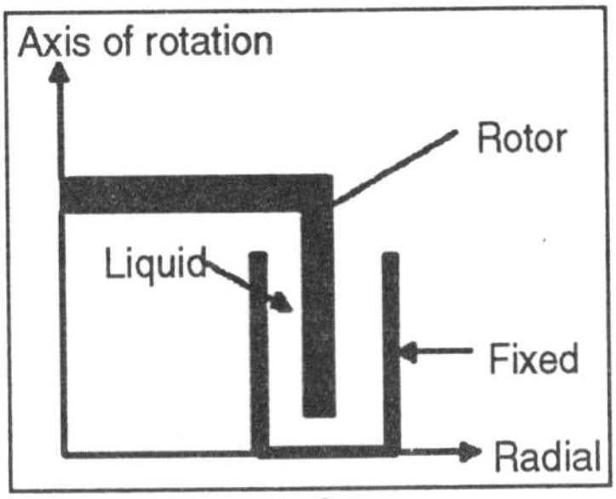

Cross-section of a double concentric cylinder. The rotor moves at a constant angular speed, driven by a TA Instruments constant stress rheometer, which supplies a known torque to the rotor. The depth of liquid is 5cm andthe diagram is approximately to scale The types of liquids investigated are varied and include complex polymers, oil drilling muds, cement paste and liquid foodstuffs.

Much of the research has been funded by industrial sponsors, as well as research councils. The common theme running through this research has been the use of sophisticated rheometric instruments to determine the characteristics of the materials under study.

Where necessary, standard theoretical results have been replaced by new solutions, instrument theory has been refined and rheometer test cells have been manufactured in-house.

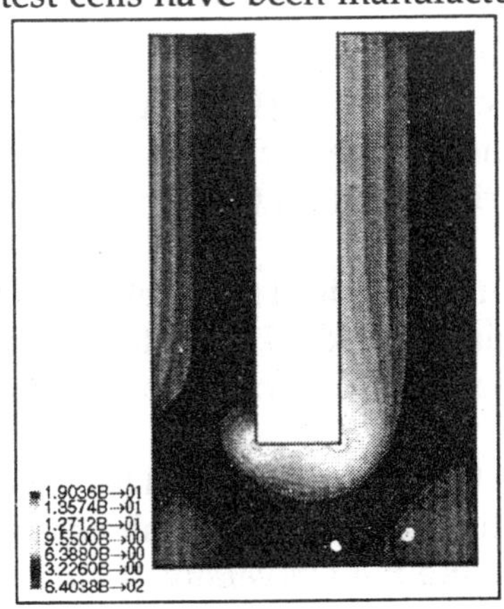

Contour plots of the shear rate in the double concentric cylinder. Note the very high shear rates that occur near the sharp corners of the rotor – here the shear rate is theoretically infinite.

The units of shear rate used are 1/s A recent example of such a test cell is depicted schematically in the top figure in which the cross-section of a 'double concentric cylinder' (DCC) geometry is shown.

This particular instrument has much wider gaps than conventional rheometer cells and requires a novel numerical solution algorithm to convert raw data to useful information.

As part of the solution, the shear rates in the gaps of the rheometer are calculated, and a contour plot depicting the results is shown in the lower figure on the right. Interestingly, the equations solved to find the shear rate are essentially those that govern the magnetic field in a recording head with a cross-section as shown on the second figure.

Previous projects in Applied Fluid Dynamics include:

- Counter current flow of liquid foodstuffs in shell and tube heat exchangers
- Headspace bubble movement in the convective processing of liquid foodstuffs
- Droplet motion in wave plate mist eliminators
- Annular two-phase flow in pipe fittings
- Cohesive sediment transport in annular flumes and estuaries
- The settling of cuttings in oil drilling fluids

In all of these studies the methodology has been to combine analytical or numerical solutions with targeted laboratory experiments to obtain a validated mathematical model.

Frequently, use is made of Computational Fluid Dynamics software especially ANSYS CFX to generate numerical solutions. We also have our own in-house software for particle tracking.

The work on two-phase flows and the wave plate demister project were both undertaken in collaboration with Professor Barry Azzopardi and the University of Nottingham.

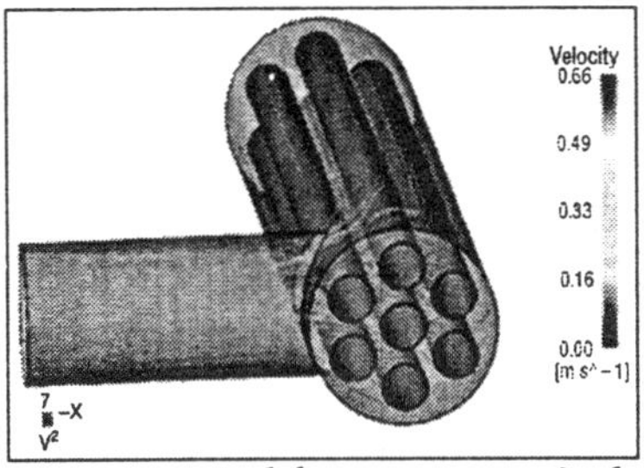

This figure was generated from a numerical simulation of the flow through part of a tubular heat exchanger, of the type used in the food processing industry. The coloured lines show the paths and the velocities of the fluid elements as they pass through the inlet section of the heat exchanger. The simulations are being used to improve the design of these types of heat exchanger so that they can be used wit more viscous food products.

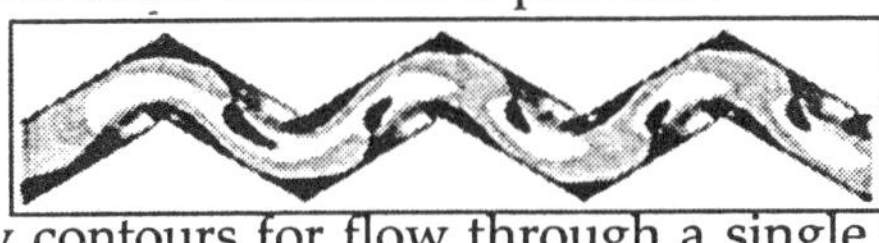

Velocity contours for flow through a single passage of a wave plate mist eliminator which has drainage channels fitted to the wave plates. Notice how the flow settles into a repeating pattern, with a jet emanating from the corner of the drainage channels

RHEOLOGY AND FLUID DYNAMICS

The final area of interest to members of the group is droplet or particle dispersion in turbulent flows. This type of problem occurs in annular two phase flow, where the smaller droplets that have been ripped off the liquid film by the core gas flow eventually redeposit through their interaction with turbulence (see below).

Annular air-water flow through a pipe contraction. In addition to the pool of water at the bottom of the pipe, droplets are ripped from the surface – particularly in the contraction itself. Such configurations are common in industrial systems.

David Graham has a particular interest in the analysis of turbulent particle dispersion. He has investigated the characteristics of dispersion statistics of inertial particles and 'turbulence modulation' effects in simple particles-laden turbulent flows.

Using in-house codes, and by referring to other experimental studies, he has checked the validity of the analyses. He is also interested in the development of particle-tracking models that work 'properly'. Finally, he is interested in the statistical behaviour of the results produced using particle-tracking models.

The results can be surprising – people have only just come to realise that the results from this kind of stochastic model are in fact variable (see below). The challenge is to characterise the variability, but without creating too many extra computations.

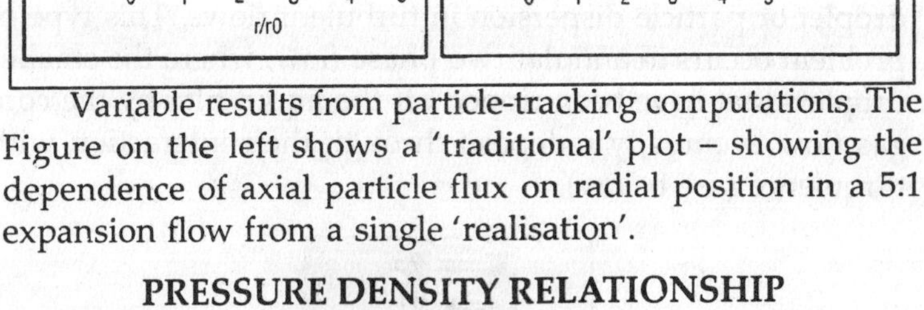

Variable results from particle-tracking computations. The Figure on the left shows a 'traditional' plot – showing the dependence of axial particle flux on radial position in a 5:1 expansion flow from a single 'realisation'.

PRESSURE DENSITY RELATIONSHIP

DENSITY AND PRESSURE

Definitions

- The density () of an object of total mass M and volume V is given by:

$$\rho = \frac{M}{V}$$

- The units are usually taken to be kg/m ^{3}.
- The specific gravity of a substance is the ratio of the density of the substance to the density of water ($\rho_{wate} = 1 \times 10^3$ kg/m^3).
- *Pressure*: When an object is immersed in a fluid or gas, it experiences a uniform inward force over its entire surface, due essentially to the weight of the fluid sitting on top of it. The pressure on the object is defined to be the force per unit area:

$$P = \left|\frac{F}{A}\right..$$

- The units of pressure are normally given in Pascals (*Pa*) *where*:

$$1\ Pa = 1\ N/m^2.$$

At the earth's surface, the pressure felt by an object due to the weight of the atmosphere above it is:

$$1\ atmosphere = 1.01 \times 10^5\ Pa.$$

The density of a material is defined as the mass per unit volume.

Density = r = mass/Volume (the Greek letter rho)

The units are = kilograms/meter^3 = kg/m^3

Its a measure of how tightly the atoms of a material are packed. It has nothing to do with the hardness of the material.

Examples:

Material	Density (kg/m^3)
air	1.29
ice	917
water	1000
aluminum	2700
lead	11300
gold	19300

The specific gravity of a material is the ratio of its density to that of water. For example, the specific gravity of aluminum would be 2.7. This number is dimensionless.

The pressure, P, is defined as the ratio of force to area:

$$P = \frac{\text{Force}}{\text{Area}}$$

The units of pressure are:

Newtons/meter^2 = N/m^2 = Pascals = Pa

Example:

A hammer supplies a force of 700 N. The hammer head has an area of 7.1×10^{-4} m^2.

What is the pressure?

P = F/A

P = (700 N/7.1×10^{-4} m^2)

P = 9.86×10^5 N/m^2

P = 9.86×10^5 Pa

VARIATION OF PRESSURE WITH DEPTH

One might guess that the deeper you go into a liquid or gas, the greater the pressure on you from the surrounding fluid will be. The reason for the increased pressure is that the deeper into a fluid you go, the more fluid, and thus the more weight, you have over top of you. We can calculate the variation of pressure with depth by considering a volume of fluid of height *h* and cross-sectional area *A* 2.

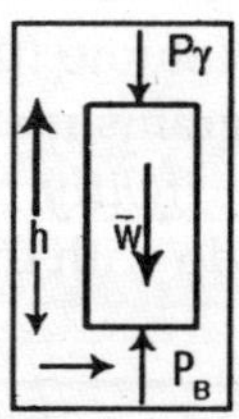

Fig. Variation of Pressure with Depth

If this volume of fluid is to be in equilibrium, the net force acting on the volume must be zero. There are three external forces acting on this volume of fluid.

These forces are:

- The force P_TA due to the pressure on top of the volume of fluid. If the fluid is open to the air, $P_T = P_O = 1.01 \times 10^5$ Pa, which is atmospheric pressure.

- The weight of the volume of fluid, w = Mg. Remembering the definition of density, $\rho = M/V$, and that the volume of the fluid may be calculated as V = Ah, we can write the weight of the fluid as $w = \rho ghA$.
- The force pushing up on the bottom of the volume of fluid, $P_B A$, due to the fluid below the volume under consideration.

If we take the up direction to be positive and add the forces we get,

$$P_B A - \rho ghA - P_T A = 0,$$

which gives,

$$P_B = P_T + \rho gh.$$

This provides the general formula relating the pressures at two different points in a fluid separated by a depth h.

Only the density of the fluid and the difference in depth affects the pressure. The shape and size of the container are irrelevant. Thus the water pressure 6 inches below the surface of the ocean is the same as it is 6 inches below the the surface of a glass of salt water.

Idea: Pascal's Principle states that any pressure applied to an enclosed fluid is transmitted undiminished to every point of the fluid.

Thus, in Fig.shown below:

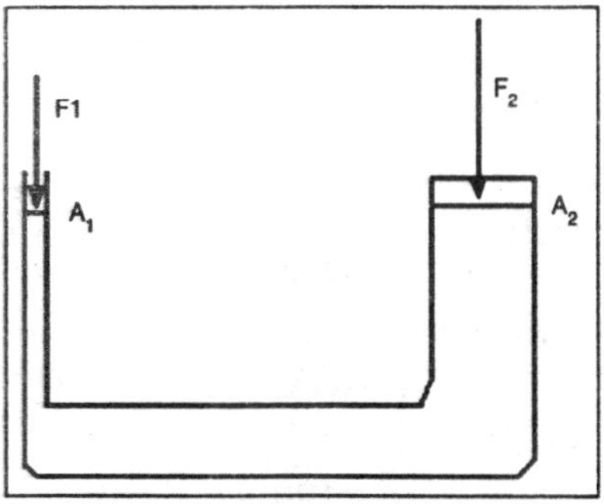

Fig. Pascal's Principlea Pressure of $P_1 = F_1/A_1$ Applied Downward to the Surface on the Left of the Container Gets Transmitted as an Equal Pressure Upward of $P_2 = P_1$ on the Surface on the other Side of the Container.

The force on the other side is therefore:

$$F_2 = P_2A_2 = F_1\frac{A_2}{A_1}.$$

Thus if $A_1 < A_2$, the transmitted force, F_2, is greater than the applied force, F_1. This is the principle behind the hydraulic press. For example, the transmitted force F_2 is used to balance the weight of a car in the hydraulic lift problem.

If a fluid is at rest, then all points at the same depth must be at the same pressure (otherwise it would be moving!). However the pressure WILL vary with depth: it will have the weight of the fluid on top of it. P = pressure, P_o = pressure due to the air (atmospheric pressure)

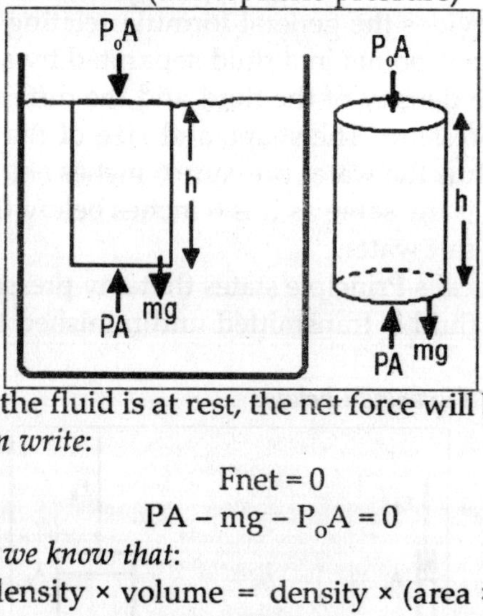

Because the fluid is at rest, the net force will be________?

So we can write:

$$Fnet = 0$$

$$PA - mg - P_oA = 0$$

However we know that:

Mass = density × volume = density × (area × height)

$$M = \rho V = \rho Ah$$

So we can substitute:

$$PA - (\rho Ah)g - P_oA = 0$$

cancel out the Area, A:

$$P - (\rho h)g - P_o = 0$$

and the pressure at any depth will be:

$$P = P_o + \rho gh$$

In words this says: The pressure at a depth 'h will be the atmospheric pressure (14.7 lbs/in^2) + (ρgh)

Example: Calculate the pressure at a depth of 3.0 m in a swimming pool.

$\rho_{water} = (1000\ kg/m^3)$

$P = P_o + rgh$

$P = (1.013 \times 10^5\ Pa) + (1000\ kg/m^3)(9.8\ m/s^2)(3.0\ m)$

$P = (1.013 \times 10^5\ Pa) + 2.94 \times 10^4\ Pa$

$P = 1.31 \times 10^5\ Pa = 131{,}000\ Pa = 19\ lb/in^2$.

at a depth of 30 m (100 ft) the pressure would be: 57 lb/in^2.
at a depth of 300 m (1000 ft) the pressure would be: 440 lb/in^2.

Example: You are a CIA operative in the jungles of Burma escaping from a rebel force. You see a stream and reeds and you cut off a reed, jump into the stream and lie about half a meter under water breathing thru the reed. Can you breathe?

Answer: lets determine the pressure on your chest...

$$P = \rho gh$$

$$P = (1000\ kg/m^3)(9.8\ m/s^2)(0.5\ m)$$

$$P = 4900\ Pa = 0.71\ psi$$

If your chest is 12 in. by 12 in. (approximately) then the area = 144 sq. in. Force = Pressure × Area = 0.7 psi × 144 sq.in. = 100 pounds.

PASCALS PRINCIPLE

Pressure applied to an enclosed fluid is transmitted equally to all points of the fluid (and therefore to all walls of the vessel containing the fluid)

QUALITIES

We can turn this phenomenon to our advantage if we alter the areas exposed to equal pressures: as in an hydraulic lift:

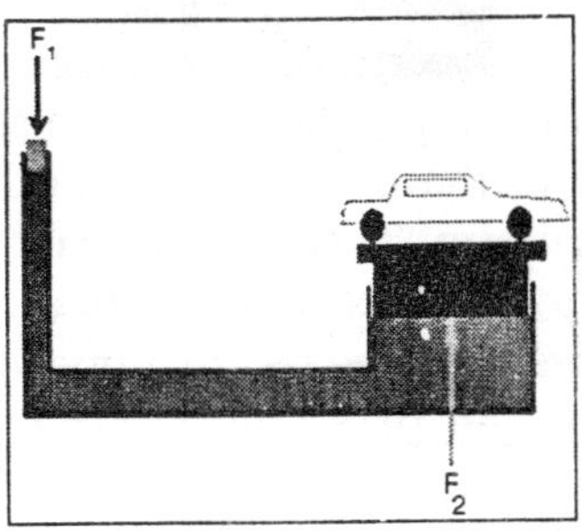

Since the pressure must be the same everywhere:

$$P_{at\ 1} = P_{at\ 2}$$

But: Pressure = Force/Area, so

$$F_1/A_1 = F_2/A_2$$

$$F_2 = (F_1/A_1)A_2$$

This says that the Force at the outlet (2) is augmented by the size of the area of the outlet. So if we make the area 1000 times larger, we can lift 1000 times the force we apply at F_1!!!

FLUID PRESSURE MEASUREMENT

Since static fluid pressure is determined by the fluid density and depth, the depth or height difference of a given liquid is commonly used for pressure measurement:

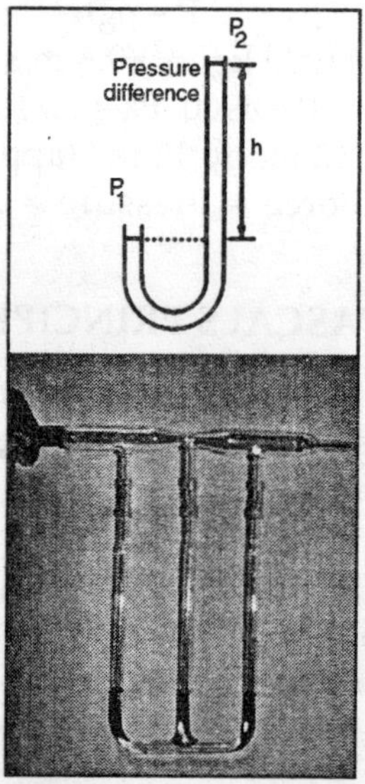

The fact that the liquid levels are the same in these three manometers shows that the pressure in the glass manifold above them is uniform. This is under static conditions with no air flow through the system so that all parts of it are atatmospheric pressure. The fact that open liquids will seek a common level is the principle behind liquid levels for construction purposes.

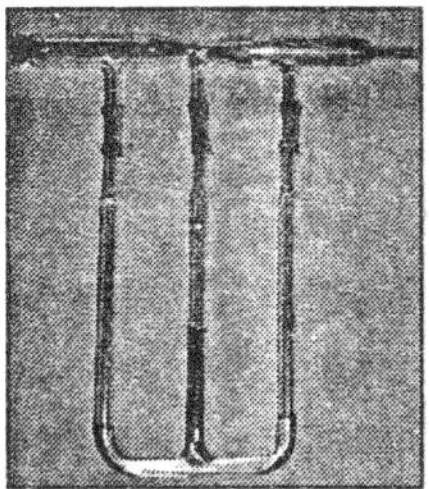

The flow in this system is from left to right, driven by a high pressure air supply. The system is said to have a positive gauge pressure exerted by the air supply. This pressure acts to push the manometer levels down, so the minimum height in the left manometer indicates that it is subjected to the highest pressure.

The fact that the centre manometer has a higher level under these conditions of rapid air flow indicates that the pressure has been lowered in the constriction by the Bernoulli effect. Note that the liquid level in the right hand tube is slightly higher than the left tube, indicating that the pressure there is slightly less than that at the left hand tube. This is indicative of the normal pressure drop in a flowing fluid from Poiseuille's law since the diameters of the top tube are the same at the left and right tubes.

RELATIONSHIP BETWEEN DEPTH AND PRESSURE

Anyone who dives under the surface of the water notices that the pressure on his eardrums at a depth of even a few feet is noticeably greater than atmospheric pressure. Careful measurements show that the pressure of a liquid is directly proportional to the depth, and for a given depth the liquid exerts the same pressure in all directions.

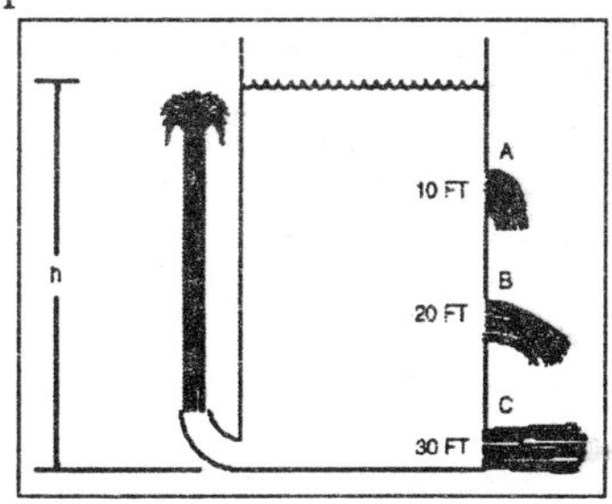

As shown in Figure above the pressure at different levels in the tank varies and this causes the fluid to leave the tank at varying velocities. Pressure was defined to be force per unit area. In the case of this tank, the force is due to the weight of the water above the point where the pressure is being determined.

Examples:

Pressure,

$$= \frac{\text{Force}}{\text{Area}}$$

$$= \frac{\text{Weight}}{\text{Area}}$$

$$P = \frac{mg}{Ag_c}$$

$$= \frac{\rho V g}{A g_c}$$

Where: m = mass in lbm

g = acceleration due to earth's gravity $32.17 \frac{\text{ft}}{\text{sec}^2}$

$$g = 32.17 \frac{\text{lbm} - \text{ft}}{\text{lbf} - \text{sec}^2}$$

A = area in ft^2

V = volume in ft^2

ρ = density of fluid in $\frac{\text{lbm}}{\text{ft}^2}$

The volume is equal to the cross-sectional area times the height (h) of liquid.

Substituting this in to the above equation yields:

$$P = \frac{\rho A h g}{a g_c}$$

$$P = \frac{\rho h g}{g_c}$$

This equation tells us that the pressure exerted by a

column of water is directly proportional to the height of the column and the density of the water and is independent of the cross-sectional area of the column. The pressure thirty feet below the surface of a one inch diameter standpipe is the same as the pressure thirty feet below the surface of a large lake.

Example: If the tank in Figure is filled with water that has a density of 62.4 lbm/ft^3, calculate the pressures at depths of 10, 20, and 30 feet.

Solution:

$$P = \frac{\rho h g}{g_c}$$

$$P_{10\text{ lbm}} = \left(62.4\frac{\text{lbm}}{\text{ft}^2}\right)(10\text{ ft})\left[\frac{32.17\dfrac{\text{ft}}{\text{sec}^2}}{32.17\dfrac{\text{lbm}-\text{ft}}{\text{lbf}-\text{sec}^2}}\right]$$

$$= 1248\frac{\text{lbf}}{\text{ft}^2}\left(\frac{1\text{ ft}^2}{144\text{ m}^2}\right)$$

$$= 8.67\frac{\text{lbf}}{\text{m}^2}$$

$$P_{10\text{ lbm}} = \left(62.4\frac{\text{lbm}}{\text{ft}^2}\right)(30\text{ ft})\left(\frac{32.17\dfrac{\text{ft}}{\text{sec}^2}}{32.17\dfrac{\text{lbm}-\text{ft}}{\text{lbf}-\text{sec}^2}}\right)$$

$$= 1872\frac{\text{lbf}}{\text{ft}^2}\left(\frac{1\text{ft}^2}{144\text{ m}^2}\right)$$

$$= 13.00\frac{\text{lbf}}{\text{m}^2}$$

Example: A cylindrical water tank 40 ft high and 20 ft in diameter is filled with water that has a density of 61.9 lbm/ft^3.

(a) What is the water pressure on the bottom of the tank?

(b) What is the average force on the bottom?

Solution:

(a) $P = \dfrac{\rho h g}{g_c}$

$$P = \left(61.9\frac{\text{lbm}}{\text{ft}^2}\right)(40\,\text{ft})\left(\frac{32.17\dfrac{\text{ft}}{\text{sec}^2}}{32.17\dfrac{\text{lbm}-\text{ft}}{\text{lbm}-\text{sec}^2}}\right)$$

$$= 2476\frac{\text{lbf}}{\text{ft}^2}\left(\frac{1\,\text{ft}^2}{144\,\text{m}^2}\right)$$

$$= 17.2\frac{\text{lbf}}{144\,\text{m}^2}$$

(b) $$\text{Pressure} = \frac{\text{Force}}{\text{Area}}$$

Force = (Pressure) (Area)
Area = pr²

$$F = 17.2\left(\frac{\text{lbf}}{\text{m}^2}\right)p(10\,\text{ft})^2\left(\frac{144\,\text{m}^2}{1\,\text{ft}^2}\right)$$

$= 7.78 \times 10^2$ lbf

MANOMETERS

A *Manometer* is a device to measure pressures. A common simple manometer consists of a U shaped tube of glass filled with some liquid. Typically the liquid is mercury because of its high density.

CASE 1

In the figure to the right we show such a *U* shaped tube filled with a liquid. Note that both ends of the tube are open to the atmosphere. Thus both points A and B are at atmospheric pressure. The two points also have the same vertical height.

CASE 2

Now the top of the tube on the left has been closed. We

imagine that there is a sample of gas in the closed end of the tube.

The right side of the tube remains open to the atmosphere. The point A, then, is at atmospheric pressure.

The point C is at the pressure of the gas in the closed end of the tube.

The point B has a pressure greater than atmospheric pressure due to the weight of the column of liquid of height h.

The point C is at the same height as B, so it has the same pressure as B. And we have already seen that this is equal to the pressure of the gas in the closed end of the tube.

Thus, in this case the pressure of the gas that is trapped in the closed end of the tube is greater than atmospheric pressure by the amount of pressure exerted by the column of liquid of height h.

CASE 3

Now we show another possible arrangement of the manometer with the top of the left side of the tube closed. Perhaps the closed end of the tube contains a sample of gas as before, or perhaps it contains a vacuum.

The point A is at atmospheric pressure.

The point C is at whatever pressure the gas in the closed end of the tube has, or if the closed end contains a vacuum the pressure is zero.

Since the point B is at the same height as point A, it must be at atmospheric pressure too.

But the pressure at B is also the sum of the pressure at C plus the pressure exerted by the weight of the column of liquid of height h in the tube.

We conclude that pressure at C, then, is less than atmospheric pressure by the amount of pressure exerted by the column of liquid of height h.

If the closed end of the tube contains a vacuum, then the pressure at point C is zero, and atmospheric pressure is equal to the pressure exerted by the weight of the column of liquid of height h. In this case, the manometer can be used as a *barometer* to measure atmospheric pressure.

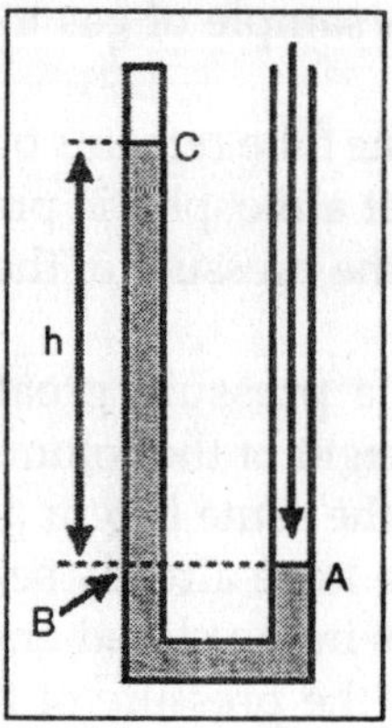

Fig. Atmospheric Pressure

We conclude with a discussion of the units for pressure measurements. Recall that pressure is defined as the force per area. The SI unit for pressure is the *pascal,* which is one *newton* per square *meter*.

For example, atmospheric pressure varies with the weather and is usually about 100 kilopascals. Another common unit for measuring atmospheric pressure is *mm of mercury,* whose value is usually about 760 mm. Put another way, if the closed end of the tube in Case 3 above contains a vacuum, the height h is about 760 mm. In many situations, measuring pressures in units of length of the liquid in the manometer is perfectly adequate. The remainder of this document discusses how to convert from those units to pascals.

So, the pressure p is:

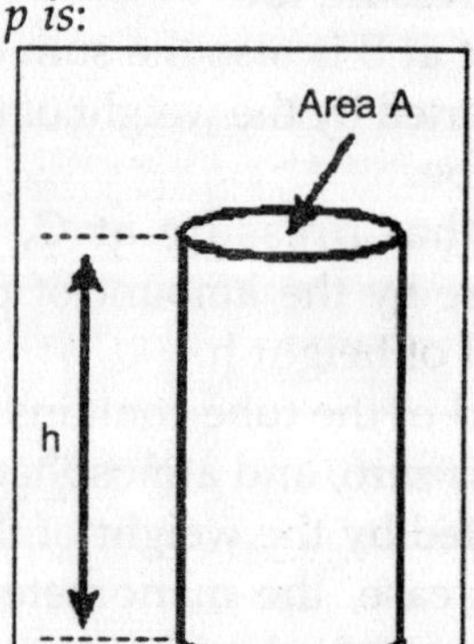

The figure to the right shows a cylinder of liquid of height h and area A.

The weight of the cylinder is its mass m times the acceleration due to gravity g.

This is the force exerted by the cylinder of liquid on whatever is just below it:

$F = m\,g$

The pressure p is this force divided by the area A of the face of the cylinder.

$$p = F/A$$

The mass of the cylinder is the density of the liquid ρ times the volume V.

$$m = \rho V$$

The volume is the area A of the face of the cylinder times its height h.

$$\begin{aligned} V &= A\,h \\ p &= F / A \\ &= mg / A \\ &= \rho V\,g / A \\ &= \rho\,Ah\,g / A \\ &= \rho\,hg \end{aligned}$$

Thus, if ρ, h and g are measured in SI units, the pressure p will be in pascals. Note that the value is independent of the area of the cylinder.

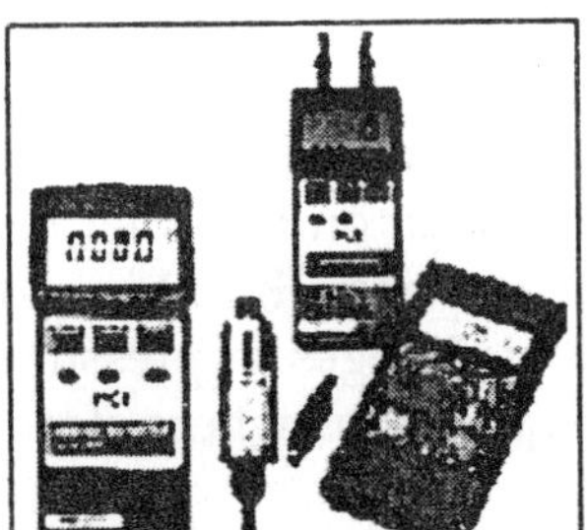

We offer manometers to determine absolute pressure, negative pressure or differential pressure in air and liquids. All our manometers are controlled by micro processors to guarantee the highest accuracy. In these manometers, they display results quickly and have an enclosure which is resistant to dust and water splash making these devices perfect for research and development. Multiple measurement ranges are available allowing you to find manometers just right for you. Three of the four series of devices have the capability to transfer data to a computer, a printer or a storage device. (Information: Manometers with an interface). There are also ISO 9000calibration certificates and components available. Our digital manometers can have a maximum value of up to 400 bar, but please contact us if you require devices with a higher maximum. If you don't find the manometers you are looking for, please contact us and we will help you find the best solution to suit your needs by calling our offices on: + 44 (0) 191 377 3398 and our technical staff will advise you regarding our measuring instruments.

Principal of manometer Manometers have a long history in the area of fluid mechanical measurements. They essentially consist of a 'U' shaped glass tube, which is filled with some liquid, typically oil, water, or mercury. At its simplest, one end of the U tube will be open to the atmosphere, while the other will be connected to whatever it is that one wishes to measure the pressure of, say a pressurized tank of gas. Manometers can also be used to make pressure measurements of liquid flows. In this case, obviously, it is important that the manometer fluid be different (and heavier!) than the liquid being measured.

Manometers are cheap, simple, and reliable. As a consequence, they are widely used, particularly in undergraduate fluid mechanical laboratories. The most serious drawback of a manometer, on the other hand, is that is has a very poor temporal response. In other words, a manometer can not pick up rapid changes in pressure. As a result, they are best suited to applications where steady-state pressure is being measured.

The analysis of a manometer is quite easy, requiring only the basic equation of hydrostatics, namely that, within a layer of constant-density fluid,

$$\frac{\rho}{\gamma} + z = \text{constant}$$

It is important to note the constant-density restriction here. One can not apply this equation across the fluid interfaces in a manometer. Rather, one must, fluid by fluid, work one's way through the manometer in order to arrive at the pressure in the test-section.

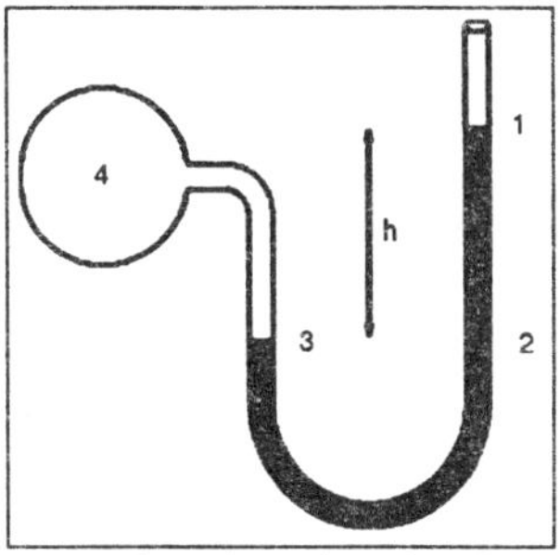

For example, consider the sketch above. The fluid in the vessel is presumed to be a gas and the fluid in the manometer is presumed to be water. The length h is the 'reading' of the manometer, in other words, the difference in elevation between the two legs.

We begin by taking p1 to be zero (gage pressure) since it is open to the atmosphere. Next, p2 is found, from (1), to be γ wh. And, since within a fluid layer there can be no horizontal variations in pressure, we see that p3 = p2! Finally, since the fluid in the reservoir is a gas (the specific weight is negligible compared to that of the manometer fluid), we immediately jump to the conclusion that p4 is also equal to p2 = γ wh. Note of caution: if the fluid in the reservoir and in the upper portion of the left-hand leg of the manometer were a liquid, we would have to consider its effect on the pressure at 4.

In the above, you can investigate the effect that the manometer fluid selection has upon the measured deflection. Note as well that a manometer can be used to measure vacuum pressures (negative P1). In this case, the deflections of the legs switch direction.

U-TUBE MANOMETER

Pressure measuring devices using liquid columns in vertical or inclined tubes are called manometers. One of the most common is the water filled u-tube manometer used to measure pressure difference in pitot or orifices located in the airflow in air handling or ventilation system.

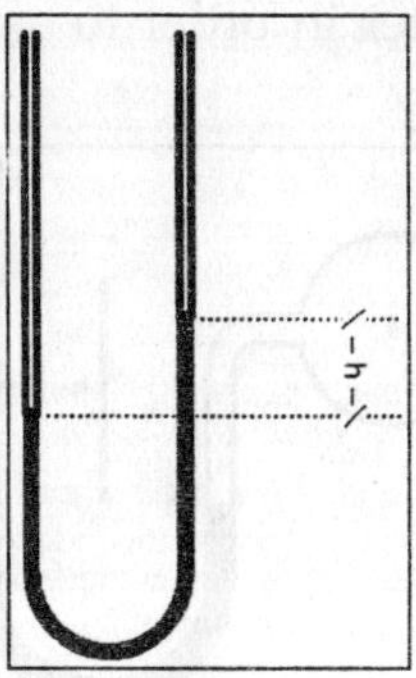

VERTICAL U-TUBE MANOMETER

The pressure difference in a vertical U-Tube manometer can be expressed as,

$$p_d = g\, h$$

where:

p_d = pressure

γ = specific weight of the fluid in the tube (kN/m^3, lb/ft^3)

h = liquid height (m, ft)

The specific weight of water, which is the most commonly used fluid in u-tube manometers, is 9.8 kN/m^3 or 62.4 lb/ft^3.

Example

A water manometer connects the upstream and downstream of an orifice located in an air flow. The difference height of the water column is *10 mm*.

The pressure difference head can then be expressed as:

$$p_d = (9.8\ kN/m^3)\ (10^3\ N/kN)\ (10\ mm)\ (10^{-3}\ m/mm)$$
$$= 98\ N/m^2\ (Pa)$$

where:

9.8 (kN/m^3) is the specific weight of water in SI-units.

INCLINED U-TUBE MANOMETER

Common problems wi[illegible]asuring pressure difference

in low velocity systems as air ventilation system are the low column heights and satisfying accurately.

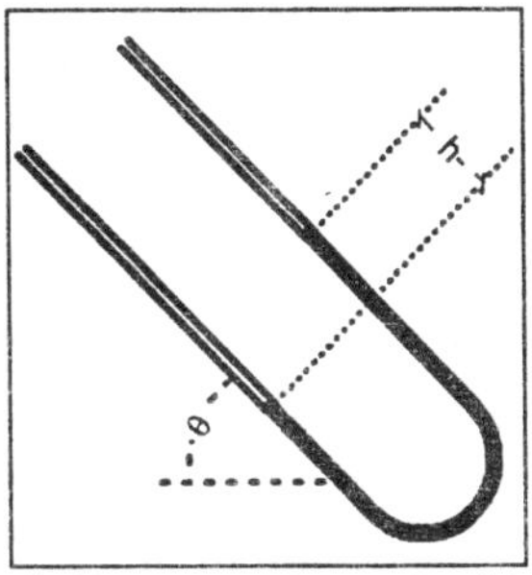

The pressure difference in a inclined u-tube can be expressed as,

$$p_d = g\ h\ \sin(q)$$

where:

θ = angle of column relative the horizontal plane

Inclining the tube manometer will increase the accuracy of the measurement.

Example: Differential Pressure Measurement with an Inclined U-Tube manometer We use the same data as in the example above, except that the U-Tube is inclined to 45°.

The pressure difference head can then be expressed as:

$$p_d = (9.8\ kN/m^3)\ (10^3\ N/kN)\ (10\ mm)\ (10^{-3}\ m/mm)\ \sin(45)$$
$$= 69.3\ N/m^2\ (Pa)$$

PRESSURE TRANSDUCERS

INTRODUCTION

A pressure transducer is a transducer that converts pressure into an analog electrical signal. Although there are various types of pressure transducers, one of the most common is the strain-gage base transducer. The conversion of pressure into an electrical signal is achieved by the physical deformation of strain gages which are bonded into the diaphragm of the pressure transducer and wired into a wheatstone bridge configuration. Pressure applied to the pressure transducer produces a deflection of the diaphragm which introduces strain to the gages. The strain will produce an electrical resistance change proportional to the pressure.

Pressure transducers are used for measuring the pressure of a fluid or air in a device or system. A pressure transducer provides to downstream display, control, or regulating devices an electrical output signal in the form of an output signal voltage or output signal current that is a measure of the pressure measured.

Pressure transducers may typically provide two types of outputs which include a cable connection for transmitting a signal representative of the sensed pressure to a remote monitoring or control device and a human-readable display for providing a local reading of the sensed pressure.

Pressure transducers that use piezoresistors are formed with a silicon substrate and an epitaxial layer, which is grown on the substrate. A portion of the substrate is removed, leaving Z thin, flexible diaphragm portion.

The piezoresistors are located in the diaphragm portion to form a pressure transducer. Capacitive pressure sensors typically include a stationary electrode having a rigid, planar conductive surface forming one plate of a substantially parallel plate capacitor and a deformable conductive member.

A generally known pressure transducer assembly for measuring fluid pressure includes at least one fluid measuring line going between the process circuit and the valve of the pressure transducer.

Tire pressure transducers are mounted on a wheel inside a tire and detect air pressure within the tire. An air pressure transducer includes a radio frequency transponder that responds to received radio frequency signals. A differential pressure transducer is used for measurement of differential pressure.

Differential pressure transducers are widely used for example in the chemical industry, food industry, aviation industry, in water treatment plants, etc. for purposes of among others flow measurement in flowing media, the control of filters, speed measurement in aircraft or water craft.

Measuring diaphragms with differential pressure transducers are frequently used to ascertain fluid flow-through.

ELECTRICAL OUTPUT

Pressure transducers are generally available with three types of electrical output; millivolt, amplified voltage and 4-20mA. Below is a summary of the outputs and when they are best used.

Millivolt Output Pressure Transducers

Transducers with millivolt output are normally the most economical pressure transducers. The output of the millivolt transducer is nominally around 30mV. The actual output is directly proportional to the pressure transducer input power or excitation. If the excitation fluctuates, the output will change also. Because of this dependence on the excitation level, regulated power supplies are suggested for use with millivolt transducers. Because the output signal is so low, the transducer should not be located in an electrically noisy environment. The distances between the transducer and the readout instrument should also be kept relatively short.

Voltage Output Pressure Transducers

Voltage output transducers include integral signal conditioning which provide a much higher output than a millivolt transducer. The output is normally 0-5Vdc or 0-10Vdc. Although model specific, the output of the transducer is not normally a direct function of excitation. This means unregulated power supplies are often sufficient as long as they fall within a specified power range. Because they have a higher level output these transducers are not as susceptible to electrical noise as millivolt transducers and can therefore be used in much more industrial environments.

4-20 mA Output Pressure Transducers

These types of transducers are also known as pressure transmitters. Since a 4-20mA signal is least affected by electrical noise and resistance in the signal wires, these transducers are best used when the signal must be transmitted long distances. It is not uncommon to use these transducers in applications where the lead wire must be 1000 feet or more.

STYLES OF PRESSURE TRANSDUCERS

PC Board Mountable Pressure Transducers

PC board mountable pressure transducers are generally compact economical pressure transducers designed to mount on an electrical PC board and be integrated into other products.

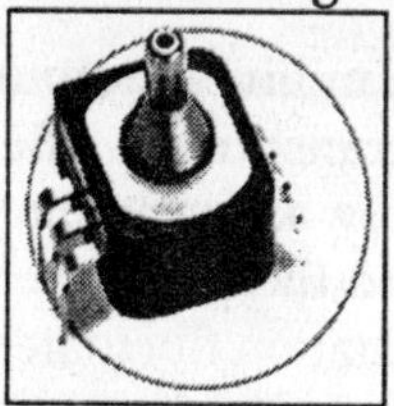

Fig. PX40 PC Board-Mo` 9untable Pressure Transcucers

General Purpose Transducers

General purpose pressure transducers are the most common since they are designed to fit the broadest set of applications.

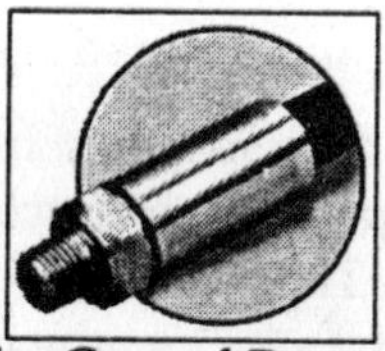

Fig. PX309 Series General Purpose Transducers

Heavy Duty/Industrial Pressure Transducers

Heavy Duty/Industrial Pressure transducers feature a much more rugged enclosure than other transducers. They are designed to accommodate heavy industrial environments. They also often feature a scalable 4-20mA output that provides much greater immunity to electrical noise which is not uncommon in industrial environments.

Fig. PX764 Industrial-Grade Pressure Transmitter

High Stability/High Accuracy Pressure Transducers

Most pressure transducers feature an accuracy of 0.25% of full scale or higher. High stability and high accuracy pressure transducers can offer errors as low as 0.05% of full scale, depending on model. Although more expensive than general purpose transducers, they may be the only option if high precision is required.

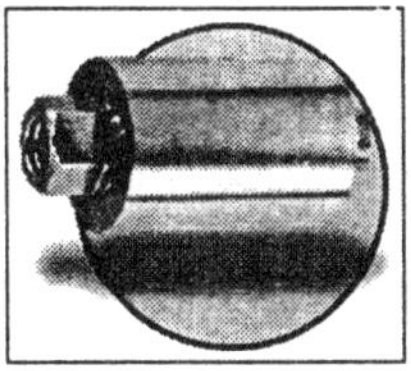

Fig. PX01 High Stability/High Accuracy Transducer

Flush Diaphragm Pressure Transducers

With flush diaphragm pressure transducers, the diaphragm is flush to the process. This eliminates a cavity above the diaphragm that could collect fluid matter from the process. In certain applications, this may be very undesirable. Those applications include monitoring the pressure of foods or liquids that have very high viscosity.

Fig. PX102 Flush Diaphragm Pressure Transducer

Special Purpose Transducers

OMEGA offers a variety of pressure transducers with special features. These include pressure transducers designed for pressure measurement in very high or low temperatures, submersible pressure transducers, barometric pressure transducers and pressure transducers with digital communications output or wireless outputs.

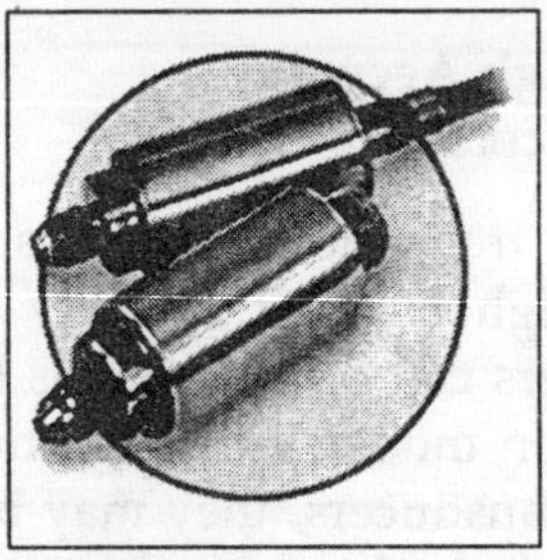

Fig. PX1004 Spcial Purpose Transducer

PRESSURE OF PLANE AND CURVED SURFACES

PRESSURE ON PLANE WINGS

As for planes, the important thing to understand is Bernoulli's Principle. What that says is that the faster air is moving, the less pressure it has. In other words, if the air is moving really fast, it doesn't push on things as hard. This is where the shape of the plane wing comes in. The plane wing is basically flat on the bottom and has a pretty big curve around the top. So the air that hits the edge of the wing gets split - some goes up over the wing and some goes down under.

The air going over has to go a lot farther than the air going under, since the top edge is curved. And it has to get there in the same amount of time. (If the air on the top took more time, then there would be an open space with no air in it, which doesn't happen.) So the air over the top actually has to go / faster/ than the air on the bottom. Since the air on the top is going faster, it has lower pressure, so it doesn't push as hard as the slower air on the bottom. So the wing gets pushed up more than it gets pushed down, and the plane goes up.

DEVELOPMENT OF CURVED SURFACES

Development is the graphical process of producing a flat pattern from drawings of a curved surface. Not all surfaces lend themselves to development; any surface with a compound curve cannot be developed. I have used development in designing a boat and in sheet metal work, but I'm sure it is useful in other fields as well.

There are three methods of development; parallel line development, radial line development, and triangulation.

Parallel Line Development

Parallel Line Development uses geometry to measure a series of heights above a base line, and marks these heights off on a series of parallel lines. Often applied to pipe, it can be applied to any object with a constant cross section, as long as a view perpendicular to the cross section is available.

Basically it is nothing more than a series of lengths along the surface of a cylinder.

This is the plan we'll start with. No, I don't know what it is. We'll develop the surface of the small branch

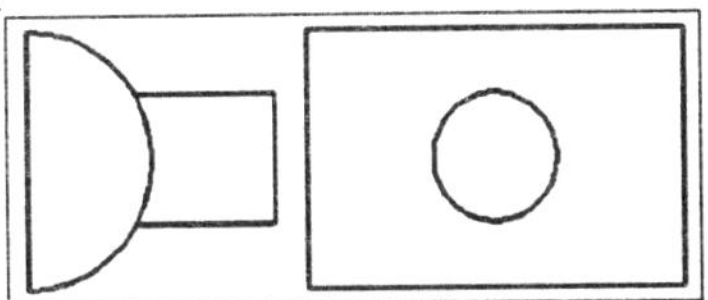

Divide the plan view of the branch of the tee with a convenient number of equally spaced radii. I've used 9 radii because it produces eighths. Because this part is symetrical, I've only marked off half of it - I could actually only mark a quarter, since it's symetrical about two axes.

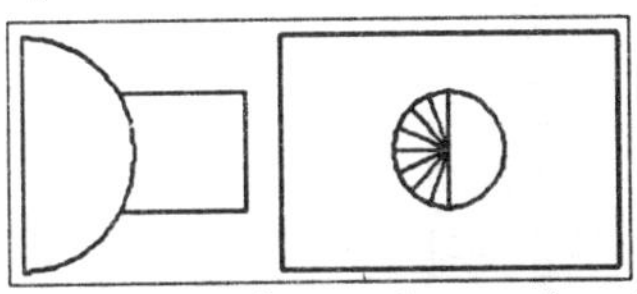

Extend perpendicular lines from the ends of each radii to the curved edge of the profile view of the branch.

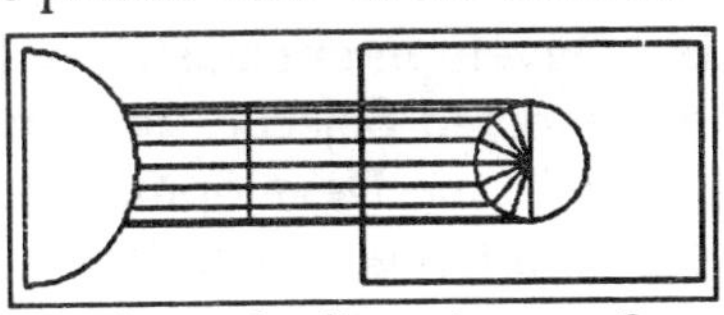

From the points where the lines in step 2 meet the curved edge of the profile view, drop vertical lines down. Add an extra line where the lines cross the base line.

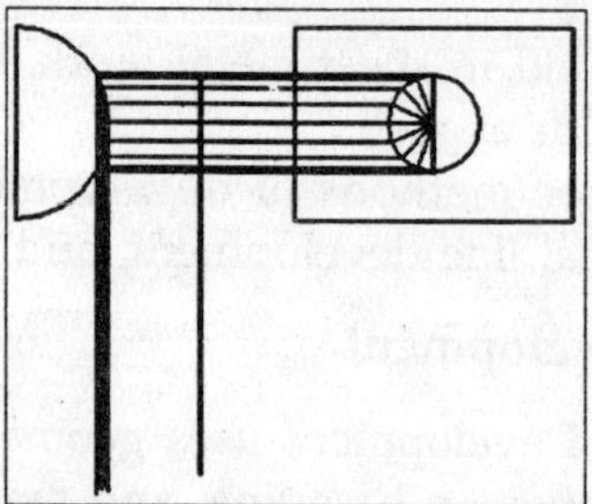

Measure the distance between the points where the radii meed the edge of the plan view, and mark off this distance on the vertical lines drawn in step 3.

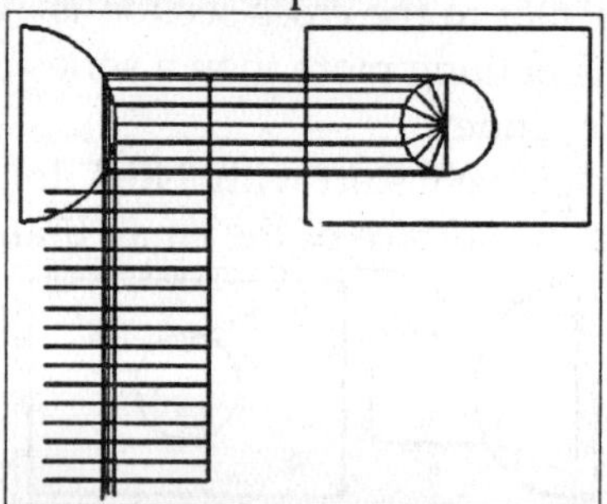

Connect the dots where the corresponding lines cross.

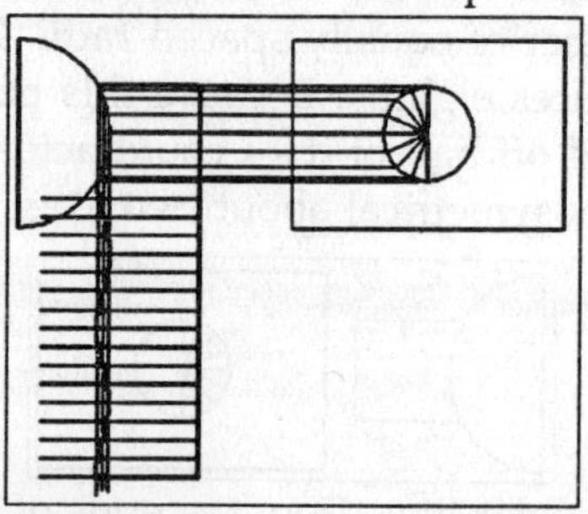

Radial Line Development

Radial Line Development basically just Parallel Line Development for tapered objects - ie. cones, pyramids, etc. Alternately, Parallel Line Development is just Radial Line Development for an infinitely small taper.

This is the plan we'll start with; a truncated cone with angle cut out. This is actually a design under consideration for the turret of my tank

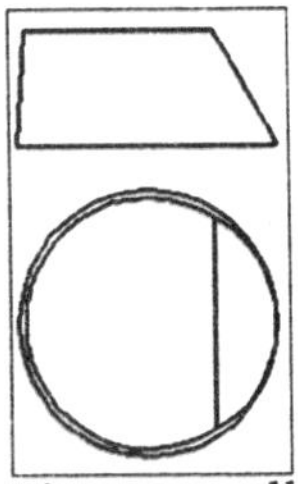

Divide the plan view into equally spaced radii. The more the better. As this object is symetrical, we only need to do this to half the plan - the other half will be a mirror image.

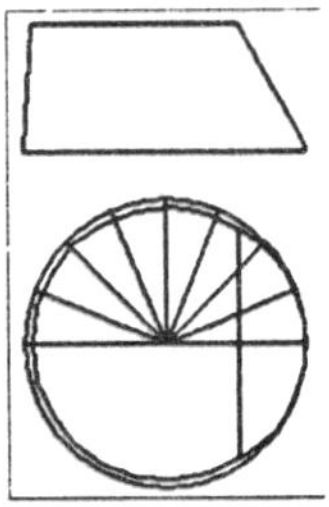

Extend vertical lines from the points where these radii meet the circumference of the plan view upward to the base line of the profile...

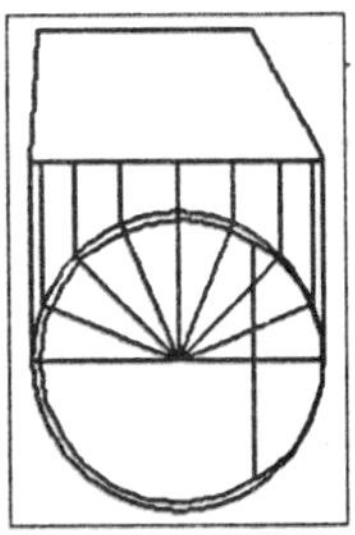

and then to the apex of the cone. This cone has a very shallow taper, so the apex is way up there.

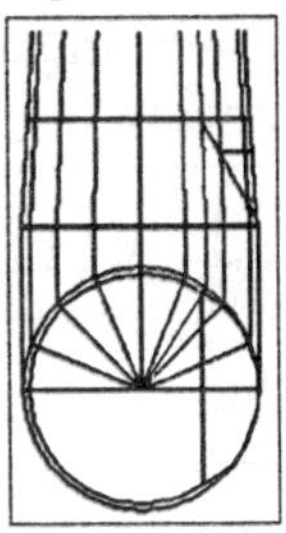

This particular design has one more point that has to be marked-the upper edge of the angled surface. Run a line from the apex through the edge of the angled surface to the base line of the profile, then vertical down to the circumference of the plan. It isn't necessary to continue the line to the centre of the plan, but it looks nice.

Draw horixontal lines from the points along the angled surface to the outermost radii on the profile view.

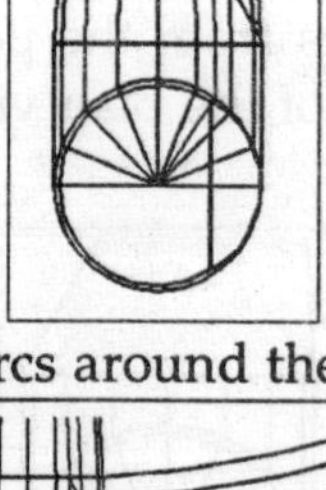

and continue them as arcs around the apex,

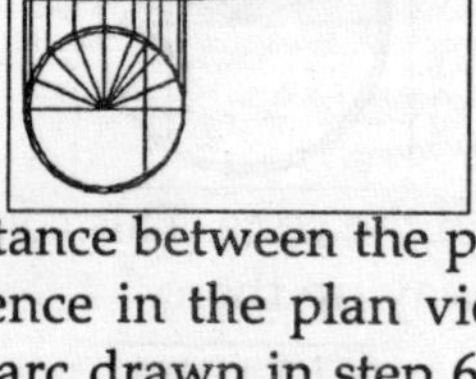

Measure the distance between the points where the radii meet the circumference in the plan view, and mark these distances off on the arc drawn in step 6 that cooresponds to the base line in the profile view

Connect the dots where the radii cross their cooresponding arcs. Remember that you've only got half a pattern here; the other side is a mirror image of this one.

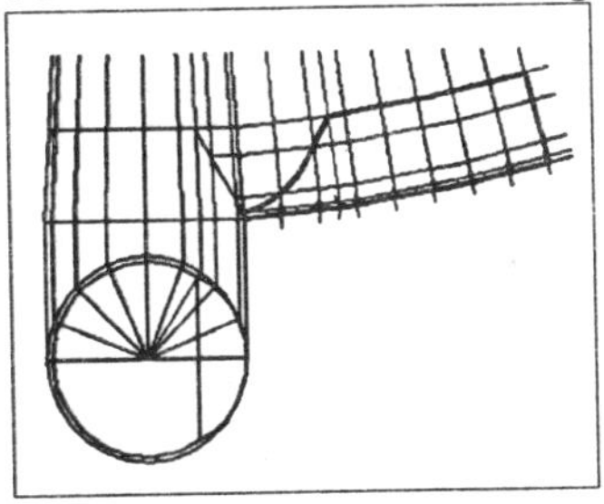

Triangulation

Triangulation is used for irregular surfaces. A good example is a boat hull. However, not all hulls can be developed-only those that do not have compound curves.

A little simple geometry is first required. The length of a line can be derived from the (X,Y,Z) coordinates of it's endpoints, using the following formula:

$$L = ((X_1 - X_2)^2 + (Y_1 - Y_2)^2 + (Z_1 - Z_2)^2)^{1/2}$$

If the distance L_{XY} in the XY plane between the endpoints is known, this formula can be simplified to:

$$L = (L_{xy}^2 + (Z_1 - Z_2)^2)^{1/2}$$

Assuming one point to be the origin simplifies this even further, and allows the Z dimension to be measured directly:

$$L = (L_{xy}^2 + Z^2)^{1/2}$$

This last formula is used to calculate the distance between two points on a surface from the distance between them on the plan drawing and the height difference on the profile drawing. Triangulation can be VERY tedious.

This is the plan we'll start with; a simple boat

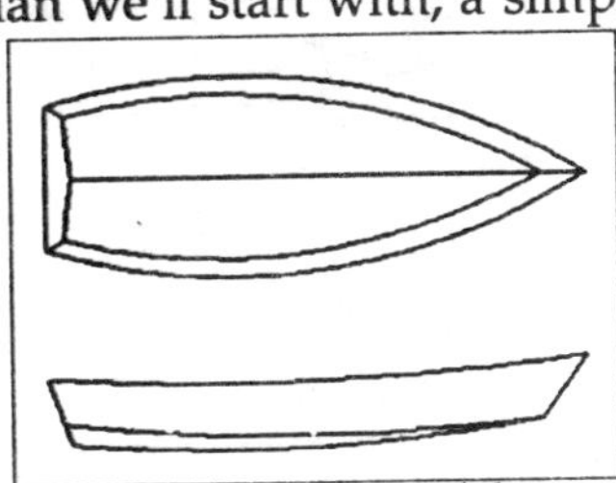

The first step is to draw vertical lines through the plan.

Izdrew nine lines; the more lines, the more accurate the developed surface. It is not necessary for the lines to coincide with the extreme front and back of the plan.

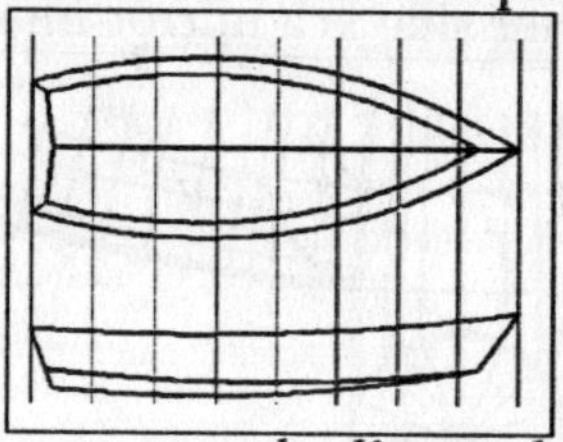

On the plan view, measure the distance between two points on the plan, and the height *of the two* same *points on the profile.*

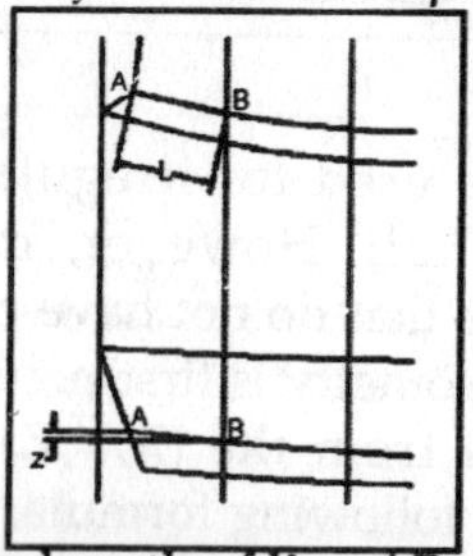

Calculate the true length of line AB using the formula $L = (L_{xy}^2 + Z^2)^{½}$. Elsewhere, draw a circle of this radius. Point A will be at the centre, and point B somewhere on the circumference of this circle.

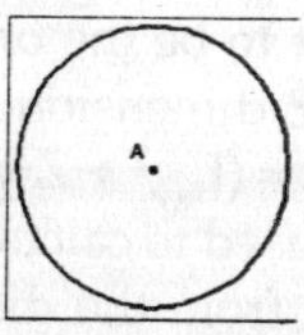

Repeat step 2 with two more points, one of which was used in step 2. In this case, I am using point B to point C.

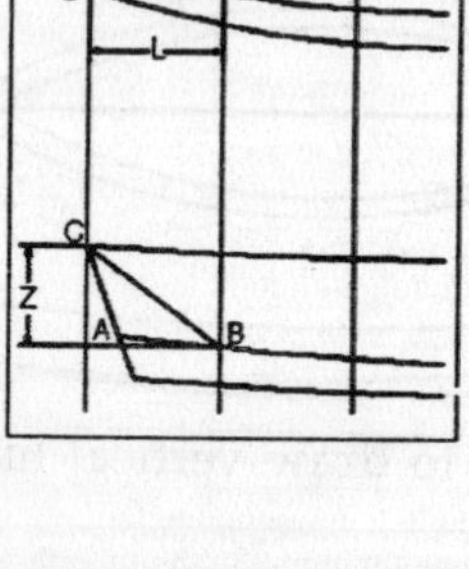

Calculate the true length of line BC as in step 3, and draw a circle of this radius at an arbitrary point on the radius of the circle drawn in step 3. As point B is at the centre of this circle, connect the centers of the two circles with line AB.

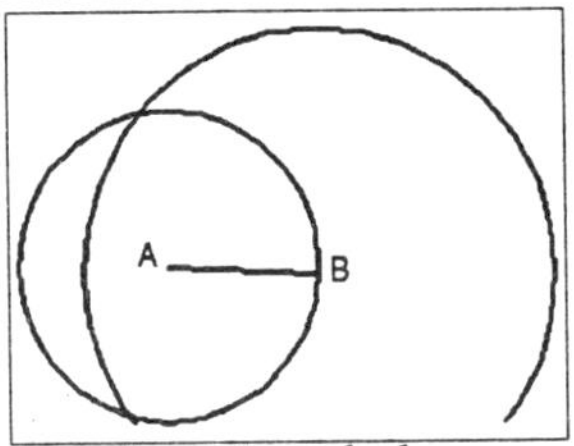

Now we complete the triangle by measuring the distance between points A and C.

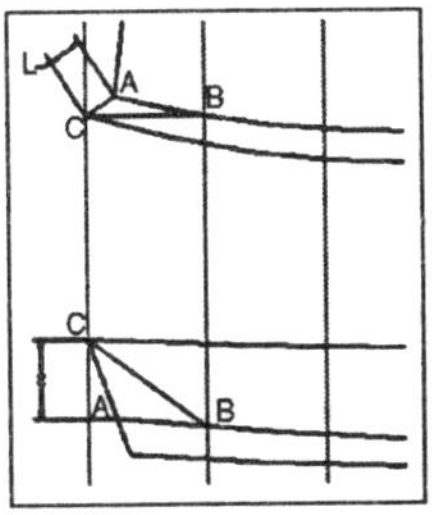

Calculate the true length and draw a circle of this radius around point A. Point C lies where the circles around points A and B intersect.

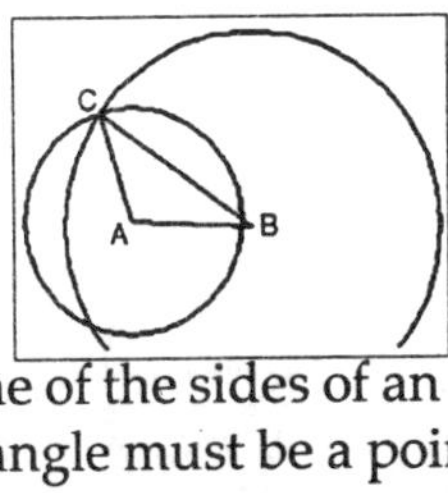

Now measure one of the sides of an adjacent triangle. One of the ends of this triangle must be a point on the first triangle; ie. points A, B or C.H Here I am measuring between points C and the new D.

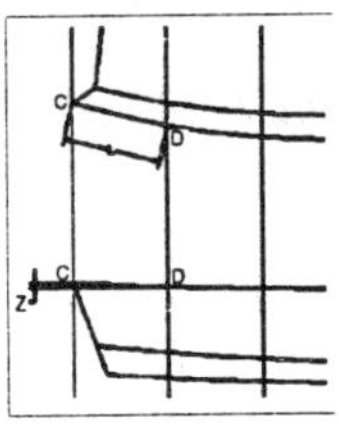

circle around point C. Point D will be on the circumference of this circle... but where?

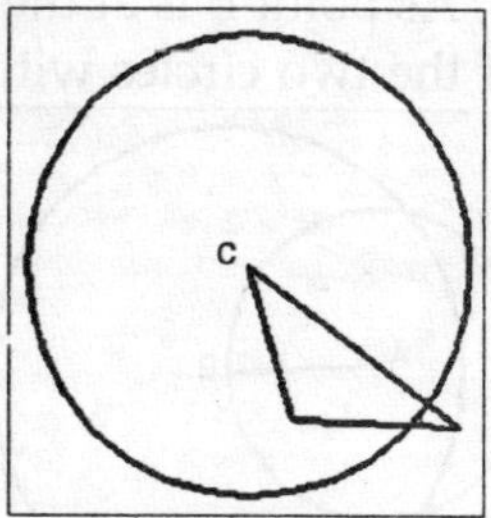

Measure the other side of the adjacent triangle.

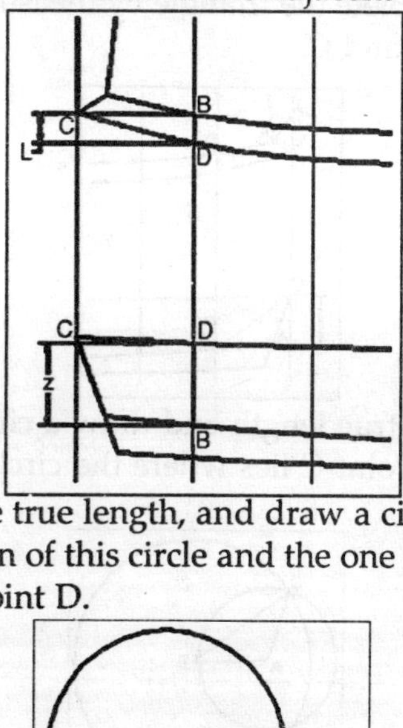

Calculate the true length, and draw a circle around point B. The intersection of this circle and the one around point C is the location of point D.

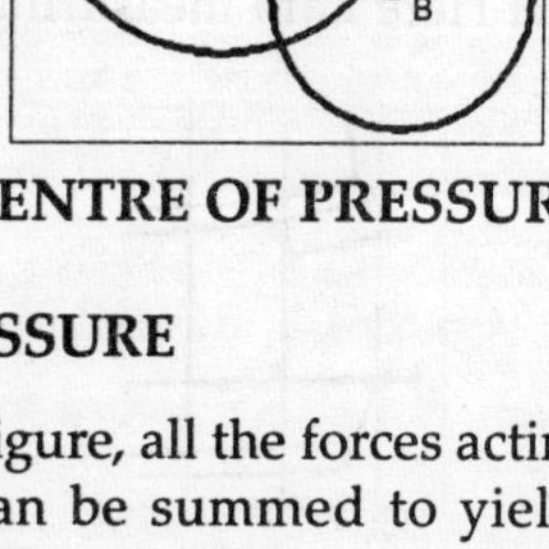

CENTRE OF PRESSURE

CENTRE OF PRESSURE

As shown in Figure, all the forces acting between the foot and the ground can be summed to yield a single ground reaction force vector (F) and a free torque vector (T_z). The point of application of the ground reaction force on the plate is the centre of pressure (CP). All the small reaction forces collectively exert on the surface of the plate at the CP.

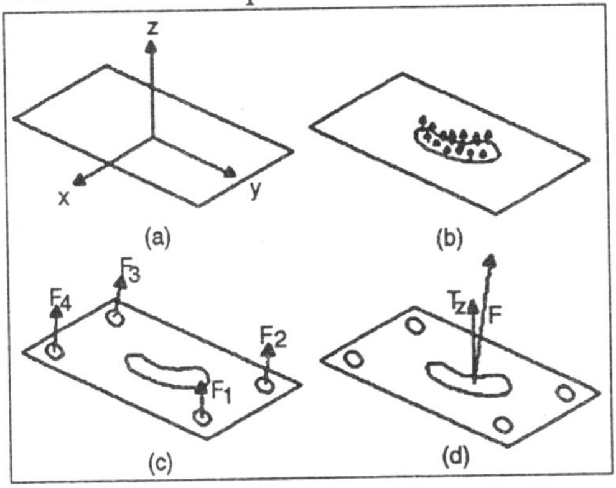

COMPUTATION OF THE CP

Generally, the true origin of the strain gauge force-plate is not at the geometric centre of the plate surface. This is due to problems in the manufacturing process. The manufacturers usually go through a series of calibrations and estimate the position of the true origin. Here, we assume that the true origin (O′ shown in Figure below) is at (a, b, c). The Z component of the CP position is always 0. The moment measured from the plate is equal to the moment caused by F about the true origin plus T_z:

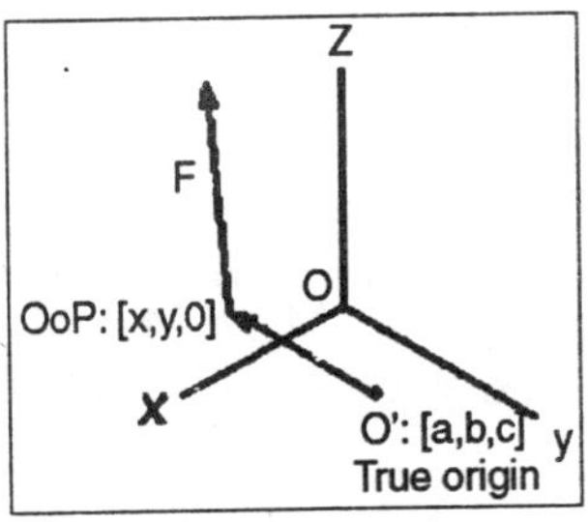

Fig. True Origin

$$M = [x - a, y - b, -c] \times [F_x, F_y, F_z] + [0, 0, T_z]$$

Therefore, the position of the CP can be computed from the moment caused by the ground reaction force about the true origin, M_x, M_y & M_z, the ground reaction force, F_x, F_y & F_z, and the location of the true origin, *a*, *b* & *c*. M_x, M_y, M_z, F_x, F_y & F_z can be directly measured from the 6 channels of the AMTI plates while the position of the true origin can be found in the calibration data sheet.

The Kistler plates provide a different channel configurations: F_{1z}, F_{2z}, F_{3z}, F_{4z}, $F_{1x} + F_{2x}$, $F_{3x} + F_{4x}$, $F_{1y} + F_{4y}$, & $F_{2y} + F_{3y}$. In Figure a, the location of the sensors are described by three distance factors: *a*, *b* & g. Among these g is the depth of the sensor centre from the surface. The sum of the moments caused by the four forces is equal to the moment caused by the ground reaction force (F) plus the free vertical torque (T_z) as shown in Figure b.

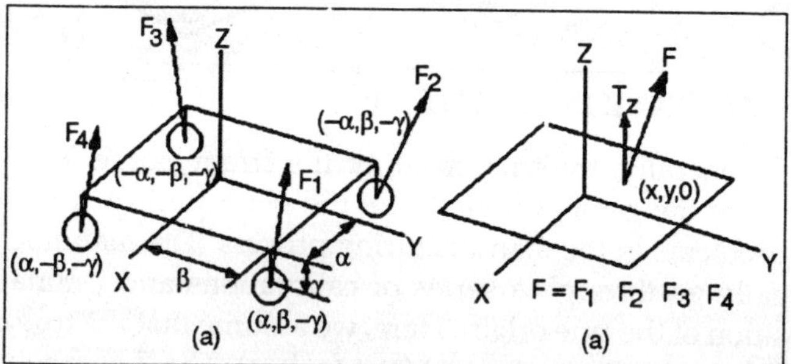

$$M_1 = (\alpha, \beta, -\gamma)\times(F_{1x}, F_{1y}, F_{1z}) = (\beta F_{1z} + \gamma F_{1y}, -\gamma F_{1x} - \alpha F_{1z}, \alpha F_{1y} - \beta F_{1x})$$

$$M_2 = (-\alpha, \beta, -\gamma)\times(F_{2x}, F_{2y}, F_{2z}) = (\beta F_{2z} + \gamma F_{2y}, -\gamma F_{2x} + \alpha F_{1z}, -\alpha F_{2y} - \beta F_{2x})$$

$$M_3 = (-\alpha, -\beta, -\gamma)\times(F_{3x}, F_{3y}, F_{3z}) = (-\beta F_{3z} + \gamma F_{3y}, -\gamma F_{3x} + \alpha F_{3z}, -\alpha F_{3y} - \beta F_{3x})$$

$$M_4 = (\alpha, -\beta, -\gamma)\times(F_{4x}, F_{4y}, F_{4z}) = (-\beta F_{4z} + \gamma F_{4y}, -\gamma F_{4x} - \alpha F_{4z}, \alpha F_{4y} + \beta F_{4x})$$

$$M = (M_x, M_y, M_z) = M_1 + M_2 + M_3 + M_4$$
$$= (x, y, 0)\times(F_x, F_y, F_z) + (0, 0, T_z)$$

APPLICATION

As shown in above equations, the CP position, (x, y), can be very sensitive to errors in the moment & force components for a small F_z. This is why we get erroneous CP positions at the beginning and end of the foot contact phase where F_z is fairly small. A small error can be easily introduced during the zero-setting procedure if the system is not well set up.

For this reason, one needs to set the threshold for F_z. If F_z is smaller than the threshold, the programme does not generate the CP coordinates.

The CP position needs to be corrected if one adds a pad on the surface of the plate and use the centre of the pad surface as the origin of the reference frame.

MODEL ROCKET CENTRE OF PRESSURE(XCP)

Input Data Defined

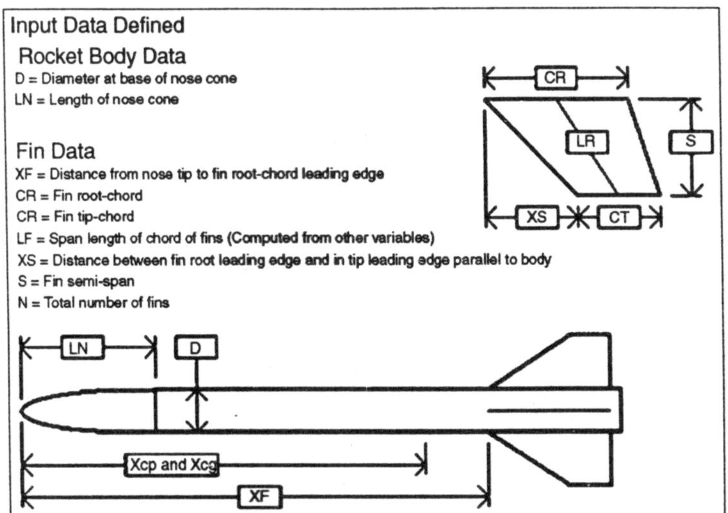

Main Stage Dimensions	
Body the diameter	D = 1.90
Length nose cone	L_N = 9.125
Distance from nose tip to fin root chord leading edge	X_F = 26.68
Fin root chord	C_R = 4.437
Sweep length of fins	X_S = 3.437
Span of fins perpendicular to fin root	S = 2.90
Total number of fins	N = 3
Rocket centre of gravity	X_{cg} = 23.25

Nose Cone

Normal Force	Centre of Pressure	Nose Cp Location	NoseCone Type
		Xn =.666 Ln	Conical Nose Shape
	Xn =.500 Ln	Parabolic Nose Shape	$C_{N\alpha-n} = 2.0$
$x_n = 466\ L_N$	Xn=.466 Ln	Ogive Nose Shape	
		Xn =.333 Ln	Ellipitcal Nose Shape

Fine/Body Interference Factor:

$$R = \frac{D}{2}$$

Span Length at Chord of Fins

$$L_F = \sqrt{S^2 + \left(X_s \frac{C_T}{2} - \frac{C_R}{2}\right)^2}$$

$$K_{fo} = 1 + \left(\frac{R}{S+R}\right)$$

$$K_{fo} = 1.24675$$

$$LF = 3.37094$$

Normal Force on n fins:

$$C_{Na\_f} = \frac{4.N.\left(\frac{S}{D}\right)^2}{1 + \sqrt{\left(\frac{2.L_F}{C_R + C_T}\right)}}$$

$$C_{Na\_f} = 10.78127$$

Normal Force on Fins in resence of Body:

$$C_{Na\_fb} = K_{fb}\, C_{N\alpha\_f}$$

$$C_{Na\_fb} = 13.44159$$

Centre of Pressure location (fins):

$$xf = XF + \frac{X_s(C_R + 2C_T)}{3(C_R + C_T)} + \frac{1}{6}\left(C_R + C_T - \frac{C_R + C_T}{C_R + C_T}\right)$$

$$X_f = 28.80654$$

Final CP Calculastion

Total Normal Force on rocket:

$$C_{Na\_T} = C_{Na\_n} + C_{Na\_fb}$$

$$C_{Na\_T} = mC_{Na\_n} + C_{Na\_fb}$$

Centre of Pressure of Entire Rocket:

$$Xcp = \frac{C_{N\alpha\_n} x_n + C_{N\alpha\_fb} x_f}{C_{N\alpha\_f}}$$

$$X_{cp} = 25.62626$$

Static Margin (Xcp an Xcg separation to body diameter ratio:

$$\frac{X_{cp} - X_{cg}}{D} = 1.250662$$

BUOYANCY

DEFINITIONS

Buoyancy is the ability of an object to float in a liquid, such as water. This concept helps to explain why some things float while other objects sink. Buoyancy is an important factor in the design of many objects and in a number of water-based activities, such as boating or scuba diving.

The mathematician Archimedes discovered much of how buoyancy works almost 2000 years ago. In his research, Archimedes discovered that an object is buoyed up by a force equal to the weight of the water displaced by the object. In other words, a inflatable boat that displaces 100 pounds (45 kilograms) of water is buoyed up by that same weight of support. An object that floats in the water is known as being positively buoyant. An object that sinks to the bottom is negatively buoyant, while an object that hovers at the same level in the water is neutrally-buoyant.

This same idea helps to determine what will float in water and what will sink. If an object weighs more than the weight of the water it displaces, it will sink. If the object weighs less, it will float. This helps explain why a heavy ship can easily float in the water, while a much smaller and lighter brick will sink quickly. It isn't the size or shape of an object that primarily determines buoyancy, but the relation between an object's weight compared to the weight of the water the object displaces.

Buoyancy is important in a surprising number of fields. Designers and engineers must design boats, ships and seaplanes in a way that ensures that they remain afloat. In the case of submarines, experts developed ways to make them sink and bring them back to the surface. Many objects were developed with buoyancy in mind, such as life preservers and pontoons. Buoyancy affects many more things than most people imagine.

Additionally, buoyancy is very important in a number of water-related sports. Many swimmers know that there are easy ways to float at the surface, such as laying on a person's back or holding a full breath. Buoyancy becomes noticeable when a swimmer tries to dive to the bottom of the pool, which can take effort. Scuba divers work with many buoyancy issues, as divers must know how to float, hover and sink in the water. In fact, scuba divers often wear extra lead weights to counteract the positive buoyancy of their bodies and gear.

If one places a copper ball in a pail of water it will sink, whereas a wooden ball will float. Whether or not a given object will sink or float in a fluid is determined by the buoyant force on the object.

The buoyant force is essentially caused by the difference between the pressure at the top of the object, which pushes it downward, and the pressure at the bottom, which pushes it upward. Since the pressure at the bottom is always greater than at the top, every object submerged in a fluid necessarily feels an upward buoyant force. Of course, objects also feel a downward force due to gravity, and the difference between the gravitational force and buoyant force on a submerged object determines whether that object will sink, or rise to the surface.

If the weight is greater than the buoyant force, the object sinks, and vice versa. It was Archimedes (supposedly while in his bath), who realized that submerged objects always displace fluid upwards (the level of water in the bathtub rose when Archimedes got in). Thus, he reasoned that the buoyant force on an object must be equal to the weight of fluid that object displaces.

If the weight of an object is greater than the weight of displaced fluid, it will sink, wherease if the weight of the object is less than the weight of displaced fluid, it will rise. Moreover, it is evident that the volume of displaced fluid is precisely equal to the volume of the submerged part of the object, so that the difference between the buoyant force and the weight is determined by the relative density of the object and the fluid. In particular, we come to Archimide's principle, which implies that

CAUSES OF BUOYANCY

Have you ever wondered why a large steel ocean liner floats, but a small steel nail sinks? The answer is surprisingly simple: The steel hull of the ship is formed in a shape that displaces a volume of water that weighs at least the same weight as the ship. If the metal used to make the ship was reshaped into a large block of steel and placed into the same water, it would sink straight to the bottom.

This is known as the principle of buoyancy. The Greek mathematician, Archimedes, was the first person to explain the phenomenon about 2000 years ago as: "Any object wholly or partly immersed in a fluid is buoyed up by a force equal to the weight of the fluid displaced by the object."

Fresh water weighs 62.4 pounds per cubic foot. If an object weighs more than the weight of the volume of water it displaces, it will sink and is said to be negatively buoyant. If an object weighs less than the weight of the volume of water it displaces, it will float and is said to be positively buoyant. If an object displaces an amount of water equal to its own weight, it will neither float nor sink, but remain suspended and is neutrally buoyant.

So you see, objects don't float on top of the water. They are held up by the force of the water beneath. Archimedes principle holds true for all objects in all fluid. However, an object with a given volume and weight will vary in buoyancy depending on the density of the material in which the object is surrounded. The denser the surrounding liquid, the more buoyant a force will be applied.

Salt water has a higher density than fresh water due to the dissolved salts, so you'll be more buoyant in salt water than in fresh water.

That's why it's easier to float in salt water. Most people are positively buoyant in either fresh or salt water. In fact, when floating motionless on top of the water, it is necessary to exhale air from the lungs in order to sink.

By exhaling, the volume of the lungs is decreased and less water is displaced resulting in less buoyancy. We can see that changing the volume of an object changes its buoyancy by changing the amount of water that it displaces.

Let's go back to our example of the ship which was reshaped into a block of steel. It is not the weight of an object that determines whether it will sink or float. It is the amount of water that it displaces as determined by the volume of the object.

When our block of metal is reformed once again into the large hull of a ship, it now displaces a much larger volume of water. If the weight of the water it displaces is greater than the weight of the ship, it will float.

Buoyancy principles are important in the design of objects that need to float, boating, scuba-diving. Scuba-divers use weight belts to offset the added buoyancy of their equipment and allows them to sink. They also use a BCD or Buoyancy Control Device to manually increase and decrease the volume of air in the device as needed to control their buoyancy. A Scuba diver's goal is to be neutrally buoyant; he wants neither to continually float to the surface nor to sink to the bottom of the ocean floor.

ARCHIMEDES PRINCIPLE

Some objects, when placed in water, float, while others sink, and still others neither float nor sink. This is a function of buoyancy. We call objects that float, positively buoyant. Objects that sink are called negatively buoyant. We refer to object that neither float nor sink as neutrally buoyant.

The idea of buoyancy was summed up by Archimedes, a Greek mathematician, in what is known as Archimedes Principle: *Any object, wholly or partly immersed in a fluid, is buoyed up by a force equal to the weight of the fluid displaced by the object.*

From this principle, we can see that whether an object floats or sinks, is based on not only its weight, but also the amount of water it displaces. That is why a very heavy ocean liner can float. It displaces a large amount of water.

Archimedes principle works for any fluid, but as divers we are mainly concerned with two different fluids: fresh water, and salt water. We need to think of fresh water and salt water as two different fluids because equal volumes of fresh water and salt water do not weigh the same. For example, a cubic foot of fresh water weighs approximately 62.4 lbs, while a cubic foot of salt water weighs approximately 64 lbs. The extra weight is because of the dissolved minerals in salt water.

Let's take a moment and look at an object in water and Archimedes Principle. If you place a 1 cubic foot object that weighs 63 lbs into fresh water, the object is displacing 62.4 lbs of water, but weighs 63 lbs. This object will be negatively buoyant - it will sink. It is however being buoyed up with a force of 62.4 lbs, so if we weighed it in the water it wouldonly weigh.6 lbs.

If we put the same object into salt water, it would still weigh 63 lbs, but would be buoyed up by a force of 64 lbs, and it would float. It would be positively buoyant in salt water. To make the object neutrally buoyant in salt water, we would have to add 1 lb of weight to the object without changing its size (without changing is displacement). Then it would weigh 64 lbs, and be buoyed up with a force of 64 lbs, thus being neutrally buoyant.

Let's expand on this and look at an example of putting

these ideas to work in a real situation. Suppose you know that a boat had lost an anchor weighing 100 lbs. Measuring a comparable anchor, we find out that the anchor displaces 1/2 cubic feet of water.

We will also assume that the anchor was lost in a fresh water lake. You do a dive and find the anchor and want to bring it to the surface but the only resource you have available are some 1 gallon milk jugs. How many would you need to tie on to the anchor to float it to the surface?

At this point we need to do a little simple math. We know that a cubic foot of fresh water weighs 62.4 lbs, so the anchor displacing 1/2 a cubic foot of water would be buoyed up with a force of 31.2 lbs. Let's round this to 31 lbs for simplicity. This means our anchor that weighs 100 lbs on land will weigh 100-31 or 69 lbs in the water. We now know we need enough 1 gallon milk jugs to generate 69 lbs of lift.

Perhaps you remember the old expression "A pint a pound the world around." This refers to the fact that a pint of water weighs about a pound. Since there are 8 pints in a gallon, we know a gallon of water must weigh about 8 lbs. Since we know a cubic foot of water weighs 62.4 lbs, this means there are about 8 gallons of water in a cubic foot. Let's put it together and solve our anchor problem.

If we need 69 pounds of lift, we divide 69 by 8 lbs per gallon to learn we need 8.625 gallons of water displacement to make the anchor neutrally buoyant. This means, we could fill 9-one gallon milk jugs with air to lift our anchor.

Let's try another. A 3 cubic foot object weighing 400 lbs is dropped into the ocean. How big of an air lift bag (in cubic feet) would you need to lift the object?

First we determine that a 3 cubic foot object in salt water would have 3x64 lbs of lift, or 192 lbs of buoyant force. If we subtract 192 from 400 we get 208 lbs. This means we need to generate 208 lbs of lift to make our object neutrally buoyant. We then divide 208 (the objects in water weight) by 64 (the weight of a cubic foot of sea water) to get 3.25 cubic feet of displacement is needed to make the object neutrally buoyant. Thus, we would need at least a 3.25 cubic foot air lift bag.

PRESSURE AND GASES

It has long been known that air has weight. The weight of the atmosphere exerts a pressure of 14.7 pounds per square inch of force at sea level. This is to say that a 1 inch column of air as tall as the atmosphere, would weigh 14.7 pounds. We commonly call this 1 Atmosphere of pressure, or 1 ATM.

Water weighs considerably more than air does, so it can exert much more pressure. It only takes a 1 inch column of sea water 33 feet tall to weigh 14.7 pounds. This means that at a depth of 33 feet deep in the ocean, there is a total pressure of 29.4 pounds per square inch (psi). This would be 2 ATMs of pressure. One ATM from the water, + one ATM from the water. We call this the ambient pressure or absolute pressure.

Every additional 33 feet of sea water, will add another 14.7 pounds of pressure, or another ATM. At 66 feet we have two ATMs of water pressure plus our 1 ATM of air pressure for an absolute pressure of 3 ATM. At 99 feet our pressure is 4 ATMs etc.

Absolute pressure differs from gauge pressure. Gauge pressure would be the pressure that would show up on a gauge at this depth. A pressure gauge would start at 0 at the surface and show 14.7 psi at our depth of 33 feet. Gauge pressure would ignore the 14.7 psi of atmospheric pressure.

And Now for a Little Math

With a little math, we can determine the ambient pressure at any given depth. Knowing that 33 feet of sea water exerts a pressure of 14.7 lbs, it is simply a matter of dividing 14.7 by 33 to determine that sea water exerts a pressure of.445 lbs per foot of depth. Fresh water being slightly lighter, requires a depth of 34 feet to equal 1 ATM, so fresh water exerts a pressure of 14.7 divided by 34, or.432 lbs per foot of depth.

Let's then find out the absolute pressure for a depth of 50 feet of sea water. To solve this, we multiply the depth, 50, times the pressure per foot,.445. This gives us an answer of 22.25. This however would be the gauge pressure at a depth of 50 feet. To find the absolute pressure, we must add in the 14.7 psi of atmospheric pressure. 22.25 + 14.7 = 36.95 Thus the pressure for a depth of 50 feet in sea water is 36.95 pounds per square inch absolute. It is common to refer to this as "PSIA", pounds per square inch absolute.

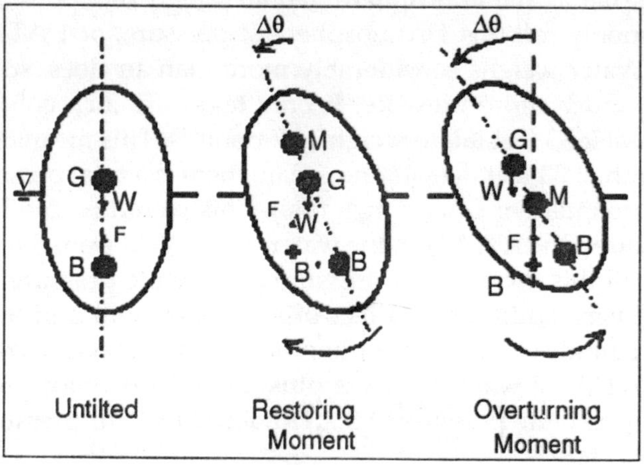

Try to figure out the gauge pressure for a depth of 75 feet in fresh water. To solve this we multiply.432 by 75 to get 32.4 psi. Since we wanted gauge pressure in this example, we do not add in the atmospheric pressure and show the pressure using psi, not psia.

STABILITY OF IMMERSED AND FLOATING BODIES

STABILITY OF FLOATING BODIES

A floating body is STABLE if, when it is displaced, it returns to equilibrium. A floating body is UNSTABLE if, when it is displaced, it moves to a new equilibrium.

Consider a floating body tilted by an angle Dq, as shown below. For the untilted body the point G is the centre of gravity of the body where the body weight, W, acts. The point B is the centre of buoyancy (the centroid of the displaced volume of fluid) where the upward buoyancy force, F_B, acts.

When the body is tilted the centre of buoyancy moves to a new position, B', because the shape of the displaced volume changes. A new point, M, may be defined, called the METACENTRE. This is the point where a vertical line drawn upwards from the new centre of buoyancy, B′, of the tilted body intersects the line of symmetry of the body. The buoyancy force, $F_{B'}$, now acts through B′.

From the centre diagram in the figure we can see that W and F_B give a RESTORING MOMENT that rotates the body back to its untilted position. From the right hand diagram in the figure we can see that W and F_B give an OVERTURNING MOMENT that rotates the body even further in the tilted direction.

Hence, we can say; if the metacentre, M, lies above the centre of gravity, G, then the body is stable. In other words the METACENTRIC HEIGHT, MG, is positive ($MG = z_M - z_G > 0$). If the metacentre, M, lies below the centre of gravity, G, then the body is unstable. In other words the metacentric height, MG, is negative ($MG < 0$).

The metacentric height, MG, is given by:

$$MG = MB - GB \quad \text{or} \quad MG = \frac{I}{Vs} - GB$$

where I is the 2nd moment of area of the plan section of the body where it cuts the waterline (this is the solid plane surface you'd see if you cut horizontally through a solid body at the water surface lifted the top part up and looked at the bottom of it!), V_S is the submerged volume (i.e. volume of fluid displaced) and GB is the distance between the centre of gravity and the centre of buoyancy ($= z_G - z_B$).

Strategy for solving buoyancy problems:

- From geometry of body and density of fluid and body equate; Weight of displaced fluid = Total weight of body. This gives the depth of immersion of the body or the weight of the body, whichever is unknown.
- To assess stability, first find the location of the centre of gravity G of the body.
- Then, find the location of the centre of buoyancy B (centroid of displaced volume). For a regularly shaped body this will be at half the height of the immersed portion of the body.

- Calculate the distance GB.
- Calculate MB, using MB = I / V_S. Note I = p $D^4/64$ for a circular section body and $bd^3/12$ for a rectangular section body (D is diameter, b and d are the sides of the rectangle).
- Calculate metacentric height, MG (= $z_M - z_G$), from MG = MB – GB. If MG > 0 then body is stable. If MG < 0 then body is unstable.

Example: A solid cylindrical pine (SG=0.5) spar buoy has a cylindrical lead (SG=11.3) weight attached as shown. Determine the equilibrium position (i.e. depth of immersion) of the buoy in seawater (SG=1.03). Calculate the metacentric height and show that the buoy is stable.

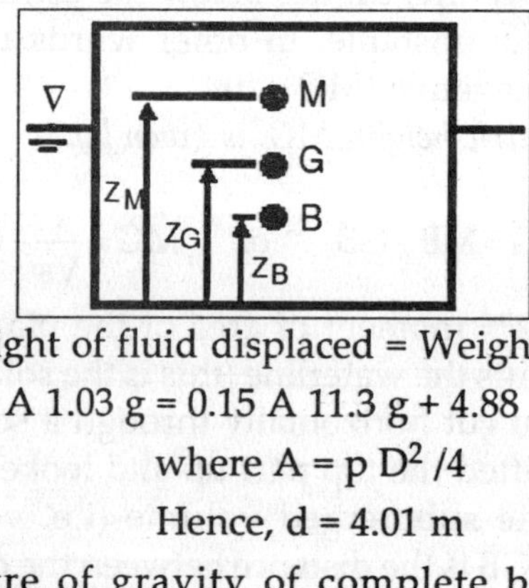

Weight of fluid displaced = Weight of body

$$d\,A\,1.03\,g = 0.15\,A\,11.3\,g + 4.88\,A\,0.5\,g$$

where A = p $D^2/4$

Hence, d = 4.01 m

Find centre of gravity of complete buoy by taking moments of the weight about the base (O):

$$GO = z_G =$$

$$\frac{0.15A11.3\gamma(0.15/2)+4.88A0.5\gamma(4.88/2+0.15)}{0.15A11.3\gamma+4.88A0.5\gamma} = 1.56\text{m}$$

$$0.15\,A\,11.3\,g + 4.88\,A\,0.5\,g$$

Centre of buoyancy (BO) will be at half the height of the submerged part of the body (i.e. half the height of the displaced volume). Hence, BO=z_B=d/2 = 2.00m.

$$GB = GO - BO = 1.56 - 2.00 = -0.44\text{ m}$$

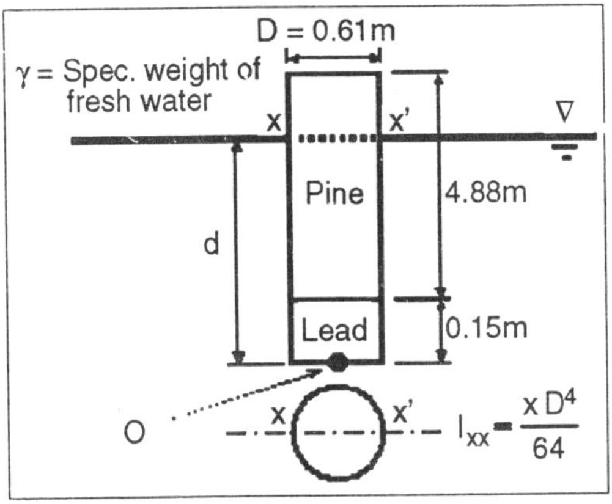

Therefore, metacentric height MG = MB – GB = 5.8×10^{-3} - -0.44 = 0.45 m . MG > 0, hence, stable.

Now, think about how much longer the pine section of the buoy would have to be in order for it to become unstable? If the buoy is stable when MG > 0 and unstable when MG < 0, we can say that the limit between being stable or unstable is when MG = 0.

In this case MB = GB. If you do the problem again, setting the length of the pine section as L (instead of 4.88m), and equating MB = GB, you can solve for L (by trial and error) to give the maximum length before instability sets in. Try it and see! [Answer = 7.20 m].

We have expanded and edited this page to provide both more theory and Worked Examples. We hope that you will find the changes an improvement.

We are always happy to receive any comments that you may have on layout or content. The new pages will be found under "Engineering/ Fluid Mechanics/ Ships and other Floating Bodies"

CENTRE OF BUOYANCY AND STABILITY

The Buoyancy Force act through the Centre of Gravity of the Displaced Fluid and is called " The Centre of Buoyancy"

There are three Types of Equilibrium:

- *Stable*: The body returns to it's original position if given a small angular displacement.

- *Neutral*: The body remains in a new position if given a small angular displacement.
- *Unstable*: The body heals further over if given a small angular displacement.

STABILITY OF FULLY SUBMERGED BODIES

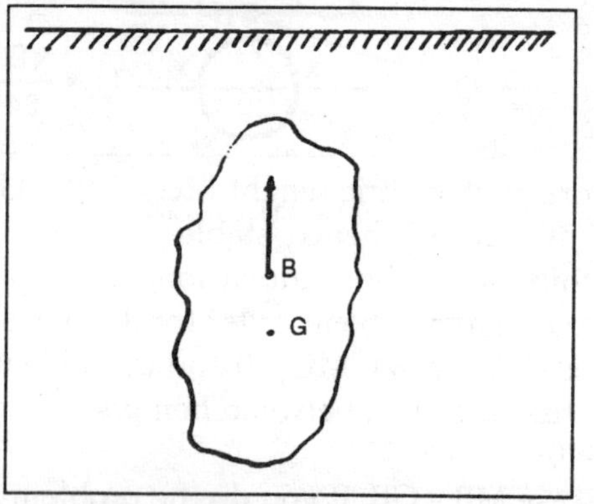

$$MB = \frac{I}{V_s} = \frac{\frac{\pi D^4}{64}}{\frac{\pi D^2}{4}} = \frac{D^2}{16d} = 5.8 \times 10^{-3} m$$

Let

- V = Volume of Body
- w = Specific weight of the fluid.
- W = Weight of the Body
- G is the Centre of Gravity and B is the centre of Buoyancy

If B and G are coincident the the Body will be in Neutral Equilibrium.

- B is below G then the Body is in Unstable equilibrium.
- B is above G then the body is in Stable equilibrium.

This last case has a righting couple of W B GWhere θ is the angle of tilt. Note that the Position of B and G relative to the Body does not change when the body is rotated.

STABILITY OF PARTIALLY SUBMERGED BODIES

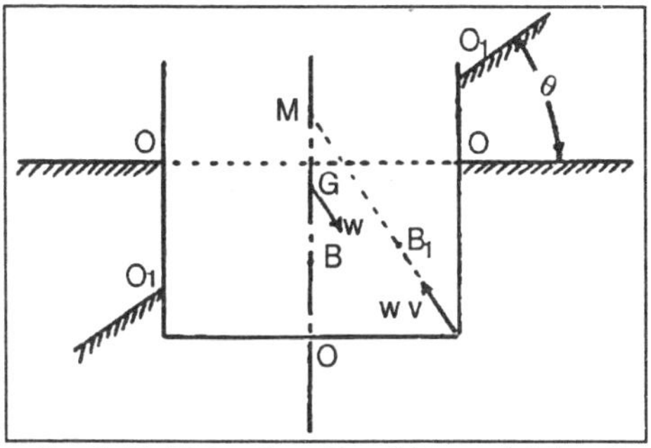

G is the C of G. B is the C. of B and The line o o is the original water surface. After tilting O′ O′ is the new water line and the angle of Tilt is θ G remains in the same position relative to the ship but the Centre of Buoyancy moves to B_1

M is the "META CENTRE" and is defined as the point where the vertical through the new Centre of Buoyancy meets the original vertical through the Centre of Gravity after a very small angle of rotation.

MG is called the metacentric height:

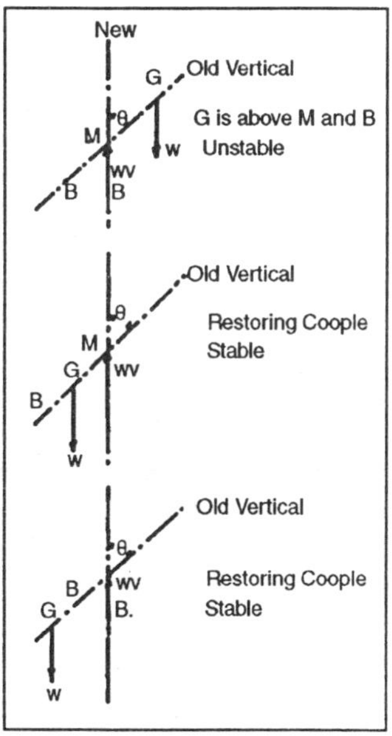

Therefore for stable equilibrium for a floating Partially Submerged Body the Metacentre must be above the Centre of Gravity G. If the Metacentric height is zero the Body will be in Neutral equilibrium. The Righting couple W GM sinθ

DETERMINATION OF METACENTRIC HEIGHT

Experimentally

Let W be the weight of the Boat plus it's Load:

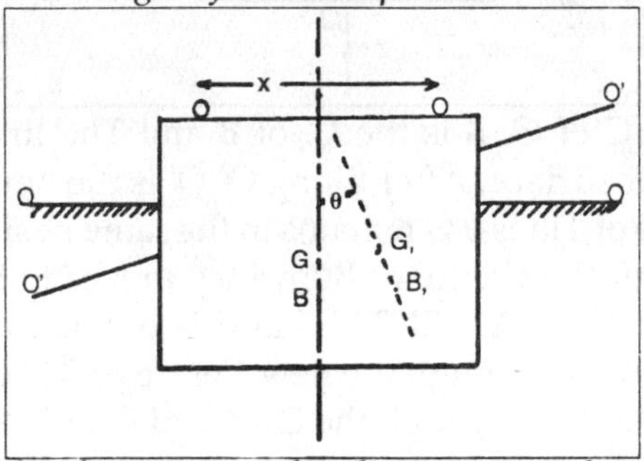

A small load w is moved a distance x and causes a tilt of angle θ. The Boat is now in a new position of equilibrium with B′ and G′ lying along the Vertical through M.

The Moment due to the movement of the load is given by:

wx = W × GM θ = moment due to movement of C of G

$$\therefore \text{GM} = \frac{w\,x}{W\theta}$$

Theory,

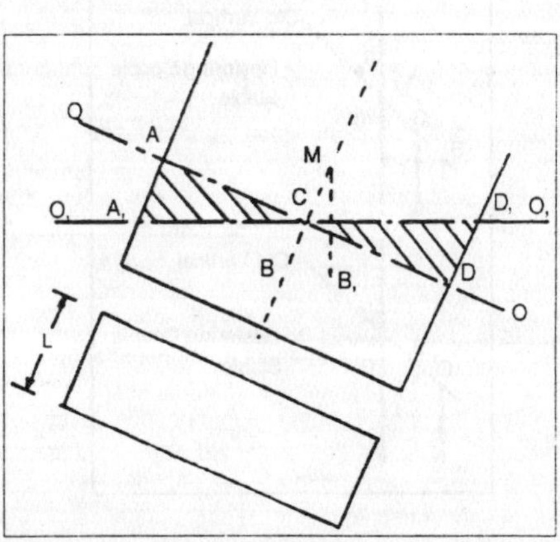

The Ship tilts from it's old waterline O O to a new waterline O′ O′ as it moves through an angle θ. Due to the movement of the wedge of water from A′ A C to D′ D C, the Centre of Buoyancy moves from B to B′

The Change in the moment of the buoyancy Force = wV X B B′

= w V × BM θ where θ is small

The Volume of the Wedge A C A′

$$=\frac{1}{2}\times\frac{b}{2}\times\frac{b}{2}q\times L=\frac{b^2L}{8}\theta$$

Therefore the Moment of the Couple due to the movement of the wedge,

$$w\times\frac{wb^2L\theta}{8}\times\frac{2}{3}b$$

$$wV\times BM\theta=\frac{wb^3L}{8}\theta$$

$$BM=\frac{b^3L}{12V}=\frac{I}{V}$$

Where,

I is the Second Moment of Area of the Water Plane Section

V is the volume of water Displaced.

Thus if the positions of B and G are known or can be calculated, then the distance GM can be determined since: 0

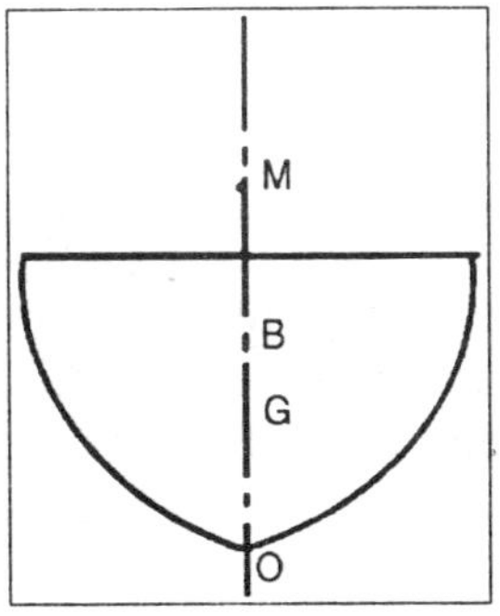

GM = BM + BG = Bm + BO – OG

There are in fact two Metacentric heights of a ship. One for Rolling and the other for Pitching. The former will always be less than the latter and unless otherwise stated, the Metacentric given will be for Rolling.

A Pontoon measuring 20ft.by 12ft. and 4ft deep, weighs 12 tons. It carries a load of 8 tons. The Pontoon sits in sea water with a density of 64 lb/cu. ft.

Metacentric Height For A Vessel With Liquid Ballast.

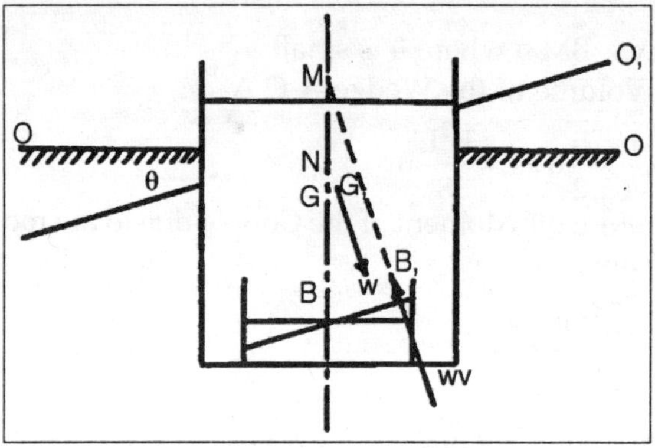

- O O is the original Waterline
- G is the Centre of Gravity of the Ship and Ballast.
- B is the Centre of Buoyancy.

The Ship is tilted through a small angle θ Clockwise giving a new waterline of O′ O′

Due to the movement of the water wedge the Centre of Buoyancy of the Ship (i.e. The C of G of the displaced liquid) moves to B′.

By the Previous,

Theory $BM = \frac{I}{V}$

Where,

I is the second Moment of Area of the Water plane area

V is the Total displacement by the Ship and it's contents.

Due to the movement of the liquid in the Ballast tank the C of G of the Ship and Ballst moves to G′. Now the vertical through G′ cuts the old vertical centre line at N. The Stability of the ship depends upon whether N is above or below M.

The Metacentric Height is now MN,

Consider the Liquid in the Ballast Tank.

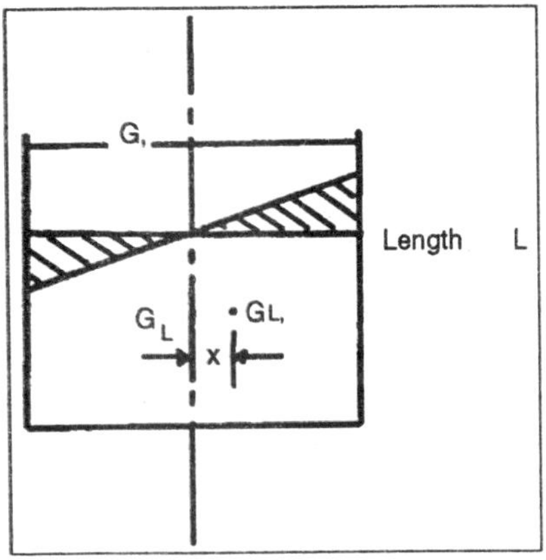

Let,

- I′ = The second moment of area of the liquid surface about it's centre line.

$$= \frac{L'B'^3}{12}$$

- V′ = the Volume of liquid in the Tank,
- w′ = the specific weight of the liquid.
- x = The movement of the C of G in the tank due to the tilting of the vessel.

The moment of the Couple due to the movement of the Wedge = The Moment of the Couple due to the movement of the C of G.

$$w'v'\ x = w' \times \frac{1}{2} \times \frac{b'}{2} \times \frac{b'\theta}{2} \times L' \times \frac{2b'}{3}$$

$$\therefore \quad V'x = \frac{L'b'^3}{12}$$

$$\therefore \quad x = \frac{I'\theta}{V'}$$

Thus the C of G of the ship and Contents moves from G to G′ where G G′ is given by:

$$W \times GG' = w'V' \times x$$

or

$$wv \times GG' = w'V' \times x$$

$$\therefore GG' = \frac{w'V'}{wV} \times \frac{I}{V'} \times \theta = \frac{w'I\theta}{wV}$$

But $GG' = GN \times \theta$

$$\therefore GN = \frac{w'I}{wV}$$

Therefore the Metacentric Height MN = MB + BO - OG - GN:

$$= \frac{I}{V} + BO - OG - \frac{w'I'}{wV}$$

Thus the Metacentric Height has been reduced by the liquid ballast by an amount,

$$\frac{w'I'}{wV}$$

Divided Ballast Tanks

If the Ballast Tank is divided into two by a longitudinal Partition, the the above equation becomes.

$$x = \frac{w_1 I_1}{wV} \text{ and } y = \frac{w_2 I_2}{wV}$$

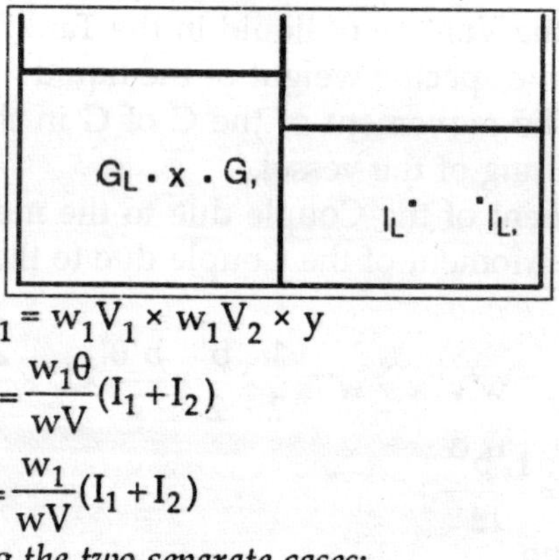

$$wV \times GG_1 = w_1 V_1 \times w_1 V_2 \times y$$

$$\therefore \quad GG_1 = \frac{w_1 \theta}{wV}(I_1 + I_2)$$

$$\therefore \quad GN = \frac{w_1}{wV}(I_1 + I_2)$$

Comparing the two separate cases:

Single Tank of width b.

$$I_1 = \frac{L_1 b_1^3}{12}$$

Same tank divided into two equal compartments.

$$I_1 + I_2 = \frac{2 \times L_1}{12} \times \left(\frac{b_1^3}{2}\right)^3 = \frac{1}{4} \times \frac{L_1 b_1^3}{12}$$

This shows that the reduction in the Metacentric height that occurs in case two is a quarter that experienced in case one. For this reason the tanks in Oil Tankers are divided into small compartments.

Right Couple With A Liquid Ballast.

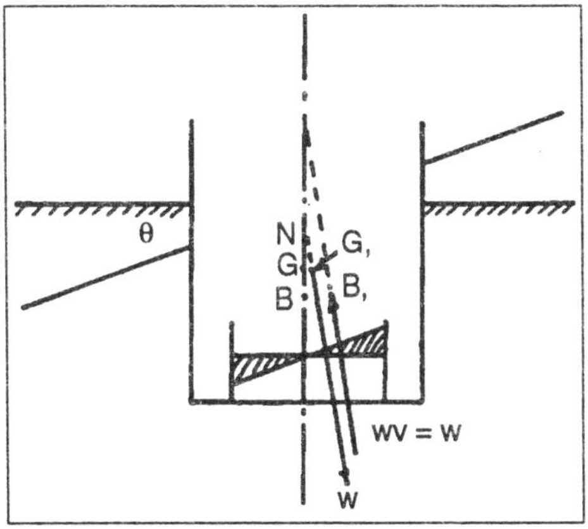

The Righting Couple due to the Buoyancy Force acting upwards through B_1 and M and the weight of the ship acting downwards through G_1 and N

$= W \times MN \times \theta$ where θ is small

This is the same as in the simple case without a liquid balance but it must be remembered that the Metacentric Height is now MN and NOT MG

Example:

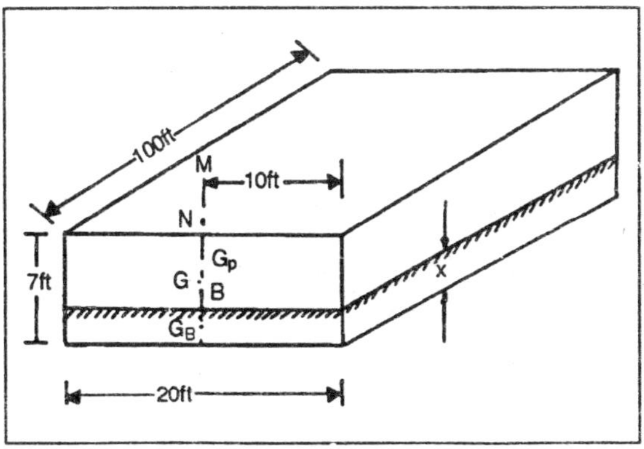

Period of Oscillation of a Floating Body

- Torque = Moment of Inertia × Angular Acceleration.

- The Solution of,

$$\frac{d^2y}{dx^2}+a^2y=0$$

is given by = A sin ax + B cos ax

For small angles of heel, the Body can be regarded as Oscillating about it's Metacentre, in a manner similar to a Pendulum about it's point of suspension.

If,

- W = The weight of the Ship
- M = The Metacentric Height
- k = The Radius of Gyration of the Ship about a horizontal Axis through the CofG.
- θ = The Rotation after a time t.

Angular acceleration,

$$=\frac{d^2\theta}{dt^2}$$

(Direction towards equilibrium positioin)

The Moment of Inertia of the ship about the C of G is,

$$\frac{w}{g}k^2$$

The Moment of Inertia of the ship about M is,

$$\frac{W}{g}(k^2+m^2)$$

Provided that m is small compared with k, these two can be regarded as equal,

$L_G \approx I_M$

For small angles of heel, the Righting Couple = W m θ

$$\therefore \text{W m q = Im} = \frac{d^2\theta}{dt^2} \approx I_G \times \frac{d^2\theta}{dt^2}$$

$$=-\frac{W}{g}k^2\frac{d^2\theta}{dt^2}$$

$$\therefore \frac{d^2\theta}{dt^2}+\frac{gm}{k^2\theta}=0$$

The Solution of this Equation is given by:

$$\theta = A\sin\sqrt{\frac{gm}{k^2}}t + b\cos\sqrt{\frac{gm}{k^2}}t$$

When t = 0 θ = 0 and therefore B = 0

When θ = o and t = T/2 i.e. T = the time of a complete oscillation.

$$A\sin\sqrt{\frac{gm}{k^2}} \times \frac{T}{2} = 0$$

Since $A \neq 0 \sin\sqrt{\frac{gm}{k^2}} \times \frac{T}{2} = 0$

The simplest solution:

$$\sqrt{\frac{gm}{k^2}} \times \frac{T}{2} = \pi$$

∴ The Periodic Time $T = 2\pi\sqrt{\frac{k^2}{mg}}$

ACCELERATION AND UNIFORM ROTATION

UNIFORM ACCELERATION

Uniform acceleration occurs when the speed of an object changes at a constant rate. The acceleration is the same over time. By relating acceleration to other variables such as speed, time and distance we are able to manipulate data in many ways.

We have a small set of very powerful formulae which can be combined to obtain results despite unknown variables. These formulae have been deduced in the Visualizing Motion section.

THE FORMULA

a = (Vf – Vi) / t

This is the formula for the average acceleration, which is our actual acceleration when dealing with uniform acceleration. Average acceleration equals to the difference in speed over time. When we have two speeds we can calculate the acceleration during the time interval, t, by subtracting the initial speed from the final speed (we get the change in speed) and then dividing by the time to get the acceleration.

Acceleration = increase in speed / time,

$d = 1/2\ (Vf + Vi) \times t$

This formula is a favourite of mine. It stems from analyzing the graph of speed V.S. time of an object's motion. Since speed = distance / time we can calculate the distance (or displacement) an object by measuring the area under the line which represents speed. The derivation of this formula is quite simple.

Using the Formulae: You now know two formulae that relate the initial speed, Vi, the final speed, Vf, the distance, d, the acceleration, a and the time, t.

When you are solving a problem where you need to find one of these variables but you are lacking another one, you should combine the two formulae to eliminate the unknown variable. Here is a list of the combinations you can use when not given a variable:

Not Given,

Formula to eliminate that variable (not needed),

d $a = (Vf - Vi) / t$

a $d = 1/2\ (Vf + Vi) \times t$

Vf $2d / t = at + 2Vi$

Vi $2d / t = -at + 2Vf$

t $2da = Vf^2 - Vi^2$

These different combinations of the same two formulae are tedious to memorize and I suggest that you remember the original formulae only. When you need to combine the formulae do so algebraically, instead of memorizing each combination.

Example: A car was travelling at a speed of 70km/h, the driver saw a rabbit on the road and slammed on the breaks. After 6.0 seconds the car came to a halt, how far did the car travel from the point where the brakes were first pressed to the point where the car stopped?

Here our knowledge of uniform acceleration is very

useful. We are not given the acceleration of the car as its stopping and we need to find the distance.

We are given:

Vi = 70km/h = 19.4 m/s

Vf = 0km/h

t = 6s

d = ?

Our formula for distance is d = 1/2 (Vf + Vi) × t

d = 1/2 (19.4 + 0) × 6

d = 58.332

The car stopped after 58 metres.

This car has very bad brakes, as 58 metres is a very long distance to come to a halt.

Lets do an example where we can combine the different formulae to get results.

Example:

- Bill jogs at 6.0km/h, he then decides to accelerate into a light run.
- Bill accelerates at 0.030km/s^2 as he runs through a distance of 40m

What is final speed?

First we convert our given information into a uniform set of magnitudes, metres and seconds.

We are given:

Vi = 6.0km/h = 1.6m/s

Vf = ?

d = 40m

a = 0.003km/s^2 = 0.30m/s^2

Lets rearrange our uniform acceleration equations to eliminate the time, which we are not given.

$2da = Vf^2 - Vi^2$

$Vf^2 = 2da + Vi^2$

$Vf^2 = 2(40)(0.3) + 1.6^2$

$Vf^2 = 26.6$

Vf = 5.2m/s

Vf = 18.5km/h

After accelerating Bill ends up running at 18 km/h.

Conclusion:

When an object is accelerating at a constant rate its motion can be modelled by two simple equations, a = (Vf - Vi) / t and d = 1/2 (Vf + Vi) × t. Using these equations gives you the ability to discover information about the motion while lacking a variable. The equations can be rearranged and substituted into each other to compensate for the lack of the distance, initial speed, final speed, acceleration or the time variable.

It is complicated to memorize every single arrangement of the two equations and I suggest that you practice creating new combinations out of the original equations. Algebraically.

Constant acceleration is very important in Projectile Motion.

It is also highly advised to visit the Visualizing Motion section.

UNIFORM ANGULAR MOTION

For example, we have already seen that motion in the presence of a uniform acceleration in one direction is described by the equations:

$$x = x_0 + v_0 t + \frac{1}{2} a t^2 \quad v = v_0 + at \quad \bar{v} = \frac{v_{FINAL} - v_{INITIAL}}{2}$$

These equations can immediately be used to tell us about angular position and angular velocity in uniformly tangentially accelerated circular motion. Recall that the angular acceleration, speed around the circle, and angular position are all related to the regular translational quantities by multiplying by the radius of the circle:

$x = r\,\theta$

$w = r\,\omega$

$a = r\,\alpha$

Therefore, after substituting these results into the equations for uniform accelerated motion, we find (after canceling all the factors of radius) that motion with uniform angular acceleration is given by:

$$\theta = \theta_0 + \omega_0 t + \frac{1}{2} \alpha^2 \; \omega = \omega_0 + \alpha \bar{\omega} = \frac{\omega_{FINAL} - \omega_{FINAL}}{2}$$

We will use this approach of taking known equation and substituting in the position, speed, and acceleration (in terms of angular quantities and radius of circle) over and over to derive new equations for the angular quantities. After a bit of practice, you can see that only ONE set of result needs to be in your head. The remainder can be derived.

TORQUE AND ANGULAR ACCELERATION

At this point, we have the kinematics of angular motion under control. Now we are ready to move on to the process of PREDICTING angular accelerations and resulting motion. Our principal goal is to determine what it is that leads to angular acceleration.

Experience shows us that forces applied to an object might, or might not lead to changes in the rotational behaviour of the object. Forces that point directly at or directly away from the axis of rotation do not lead to changes in the rotation. However, forces that point so that the direction of the force misses the axis CAN AND DO change the rotation. Such forces are said to have a *lever arm*. Lever arm is defined by drawing a line in the direction of the force, and measuring the distance of closest approach this line makes to the axis:

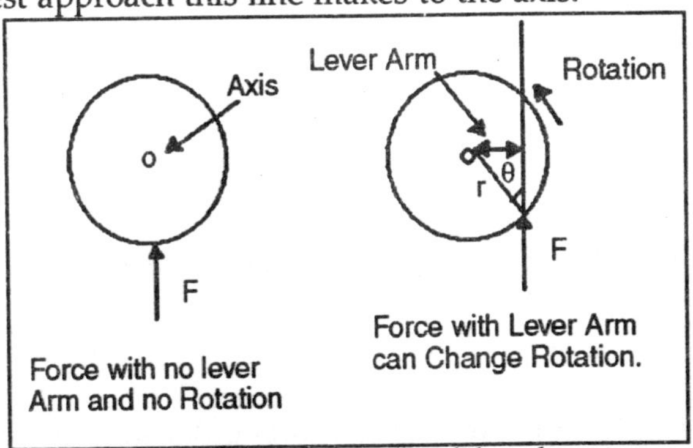

We refer to the force multiplied by the perpendicular distance to the axis, *torque* (rhymes with dork).

$\tau = Fr \sin \theta$

Torque is the quantity that changes rotational motion. For rotations, it is the object that replaces force as the cause of angular acceleration.

- Do CT8-11 the Mazur wrench problem.
- Do CT8-5 the Mazur door problem.
- Do CT8-9 on the torque of a mass on rotating rod.

NEWTON'S LAWS FOR ROTATIONS

As we have seen, torque is the quantity that changes rotational motion. Now we need to derive a precise rule for calculating exactly HOW the torque can be used to predict angular acceleration.

The first thing we state is a version of Newton's 1st Law:

Things rotating remain in their state of uniform rotational velocity unless acted on by net torque.

Similarly, once we observe angular acceleration, we know that net torque is acting. We need a rule to relate net torque to angular acceleration.

We start with Newton's 2nd Law:

F = ma

Now, imagine a point mass rotating around a circle and think about the tangential acceleration:

$F_{TENGENTIAL^{r}} = ma_{TANGEBTIAL^{r}}$

Now multiply each side by the radius of the circle:

$F_{TENGENTIAL^{r}} = ma_{TANGEBTIAL^{r}}$

The left side is just the torque. The right side has the tangential acceleration, which is just the angular acceleration multiplied by the radius of the circle:

$\tau = mr^2\alpha$

We define a new quantity that depends upon the mass and location of the mass from the axis of rotation. We call it the *moment of inertia.*

$$1 = \sum_{PARTS} m_i r_i^2$$

Therefore, net torque leads to angular acceleration, but we need the moment of inertia to tell us how large an acceleration results. Moment of inertia in rotational problems takes the place that mass has in translational problems.

- Do CT8-6 on moment of inertia of rods.
- Do CT8-7 on moment of inertia of balls on rods.

- Do CT8-8 on moment of inertia of a disk with holes.

Now put all the pieces together:

- Do CT8-10 torque and moment of inertia of bike-wheel: Torque problem
- Do CT8-14 on the torque a falling mass places on a pulley.

ROTATIONAL KINETIC ENERGY

Now, we move on to develop the ideas of energy for rotational motion. We have already seen that particles have kinetic energy due to motion. However, the speed of a given particle is due to combination of the centre of mass motion and the rotations. Therefore, the 1/2mv^2 has two pieces.

Therefore, the kinetic energy of an object that is both translating and rotating about an axis is:

$$KE = \frac{1}{2}mv^2_{CM\,TRANLATION} + KE_{ROTATION}$$

What IS the kinetic energy of rotation? Again, think about two point masses on opposite sides of a circle of radius, r, and rotating about the circle:

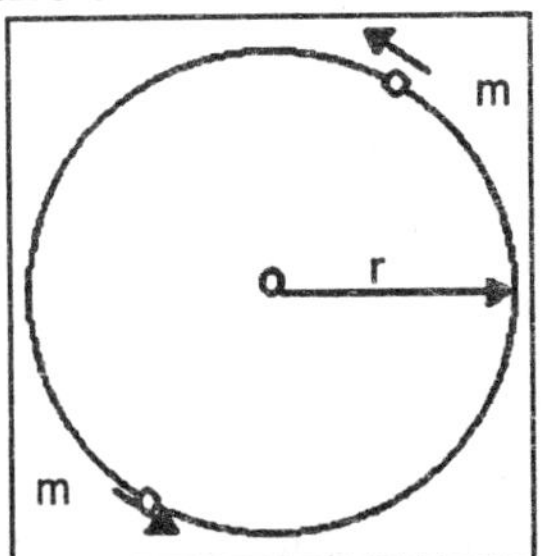

The centre of mass of this object is at the centre of the circle and is NOT translating. Therefore, the KE of this system is entirely rotational. Each mass has the same speed.

Therefore, the total KE is:

$$KE = \frac{1}{2}mv^2 + \frac{1}{2}mv^2$$

$$= \frac{1}{2}mr^2\omega^2 + \frac{1}{2}mr^2\omega^2$$

$$= \frac{1}{2}\left(mr^2 + mr^2\right)\omega^2$$

$$=\frac{1}{2}\sum_{PARTS} m_i r_i^2 \omega^2$$

$$=\frac{1}{2}I\omega^2$$

Therefore, the total KE for an object is:

$$KE=\frac{1}{2}mv_{CM}^2+\frac{1}{2}I\omega^2$$

When doing energy problems, conversion of potential energy to KE must include the correct kinetic INCLUDING rotational kinetic.

Example: Rolling without slipping

Conservation of energy is as effective in rotational problems as it is in linear motion. Many times, we can avoid nasty torque calculations by using energy conservation. As an example of the use of rotational kinetic energy, let's discuss the process of rolling without slipping.

Here is a picture of a rolling object:

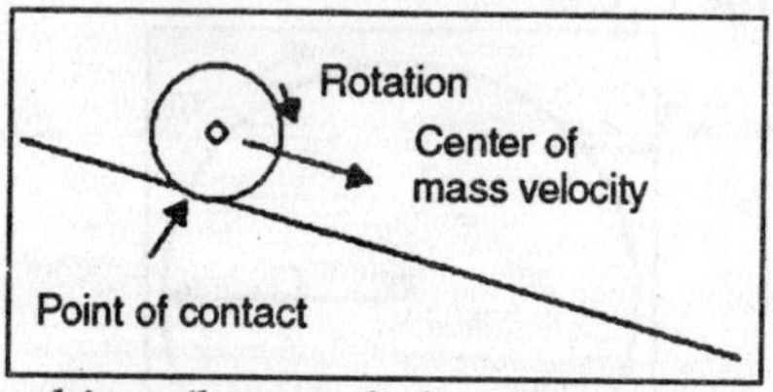

A circular object (hoop, disk, sphere, etc) can roll down an inclined plane. If the point of contact does not slip, then the motion of the centre of the circle and the motion of the circumference are related by:

$$v_{CM} = r\omega$$

Therefore, the total KE is given by:

$$KE=\frac{1}{2}mv^2CM+\frac{1}{2}I\left(\frac{v_{CM}}{r}\right)^2$$

$$=\frac{1}{2}\left(\frac{m+I}{r^2}\right)v_{CM}^2$$

Therefore, depending upon the moment of inertia, we find that different circular objects reach the bottom of the hill with different centre of mass velocities.

Chapter 2

Fluid Flow

INTRODUCTION

Unless otherwise noted, the types of fluid flows we will consider are those that are:

NON VISCOUS

There is no loss of energy due to internal friction between adjacent layers of the fluid.

STEADY STATE

At any point in the fluid, the velocity and pressure do not change with time — they have reached steady state equilibrium.

NON-TURBULENT

The fluid has no net angular momentum about any point in the fluid — there are no eddies in the flow. We usually call this type of flow to be a laminar flow. These conditions are rarely encountered in the real world. For example, water flowing through a pipe will always have some drag force on it at the pipe boundary where the water is in contact with the wall of the pipe. None-the-less, the above condition can be and are a good first order approximation. A good course in Fluid Mechanics would not neglect the effects.

TYPES OF FLUID FLOW METERS

Differential Pressure Flowmeters

In a differential pressure drop device the flow is calculated by measuring the pressure drop over an obstructions inserted in the flow. The differential pressure flowmeter is based on the Bernoullis Equation, where the pressure drop and the further measured signal is a function of the square flow speed.

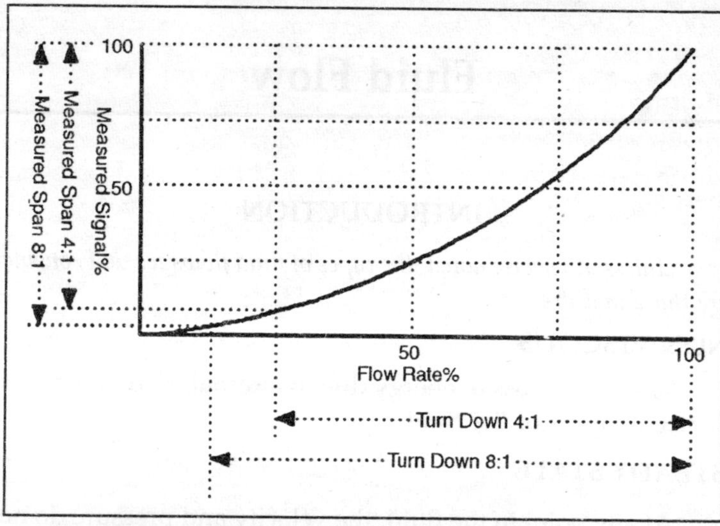

The most common types of differential pressure flowmeters are:

Orifice Plate

With an orifice plate, the fluid flow is measured through the difference in pressure from the upstream side to the downstream side of a partially obstructed pipe. The plate obstructing the flow offers a precisely measured obstruction that narrows the pipe and forces the flowing fluid to constrict.

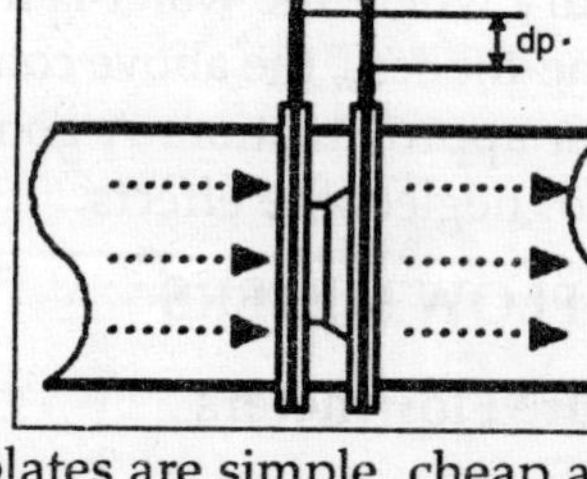

The orifice plates are simple, cheap and can be delivered for almost any application in any material.

The Turn Down Rate for orifice plates are less than 5:1. Their accuracy are poor at low flow rates. A high accuracy depend on an orifice plate in good shape, with a sharp edge to the upstream side. Wear reduces the accuracy.

- Orifice, Nozzle and Venturi Meters

VENTURI TUBE

Due to simplicity and dependability, the Venturi tube flowmeter is often used in applications where it's necessary with higher TurnDown Rates, or lower pressure drops, than the orifice plate can provide. In the Venturi Tube the fluid flowrate is measured by reducing the cross sectional flow area in the flow path, generating a pressure difference. After the constricted area, the fluid is passes through a pressure recovery exit section, where up to 80% of the differential pressure generated at the constricted area, is recovered.

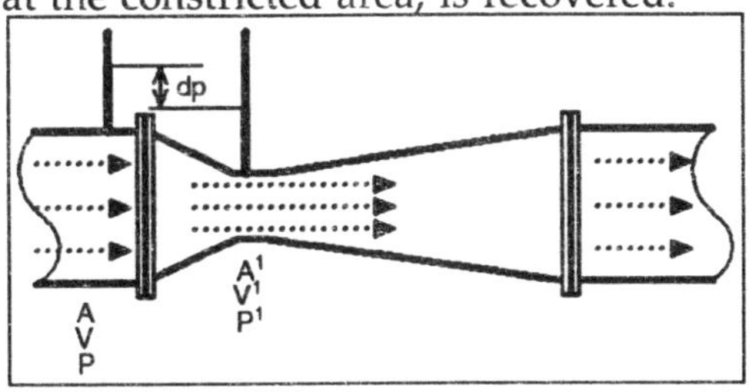

With proper instrumentation and flow calibrating, the Venturi Tube flowrate can be reduced to about 10% of its full scale range with proper accuracy. This provides a TurnDown Rate 10:1.

Flow Nozzles

Flow nozzles are often used as measuring elements for air and gas flow in industrial applications.

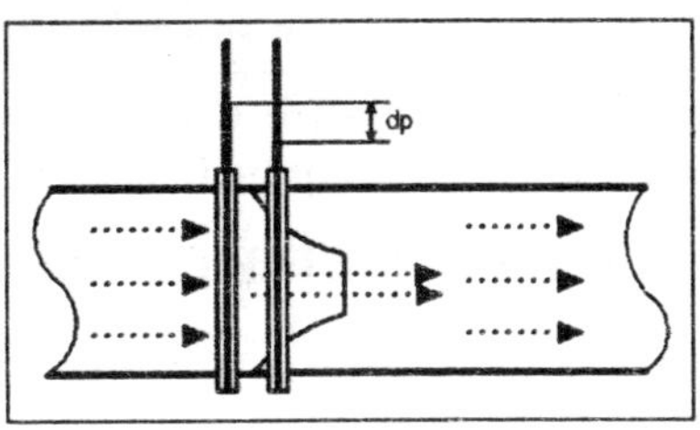

The flow nozzle is relative simple and cheap, and available for many applications in many materials.

The TurnDown Rate and accuracy can be compared with the orifice plate.

- Orifice, Nozzle and Venturi Meters

The Sonic Nozzle

When a gas accelerate through a nozzle, the velocity increase and the pressure and the gas density decrease. The maximum velocity is achieved at the throat, the minimum area, where it breaks Mach 1 or sonic. At this point it's not possible to increase the flow by lowering the downstream pressure. The flow is choked.

This situation is used in many control systems to maintain fixed, accurate, repeatable gas flow rates unaffected by the downstream pressure.

Nozzles and Venturi Meters

After the pressure difference has been generated in the differential pressure flow meter, the fluid pass through the pressure recovery exit section, where the differential pressure generated at the constricted area is partly recovered.

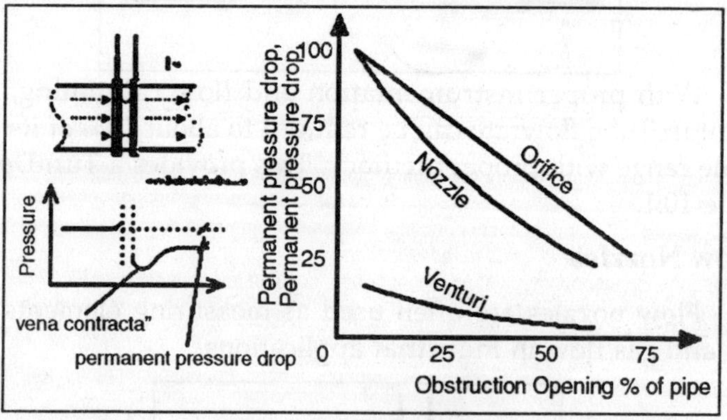

As we can see, the pressure drop in orifice plates are significant higher than in the venturi tubes.

VARIABLE AREA FLOWMETER OR ROTAMETER

The rotameter consists of a vertically oriented glass (or

plastic) tube with a larger end at the top, and a metering float which is free to move within the tube. Fluid flow causes the float to rise in the tube as the upward pressure differential and buoyancy of the fluid overcome the effect of gravity.

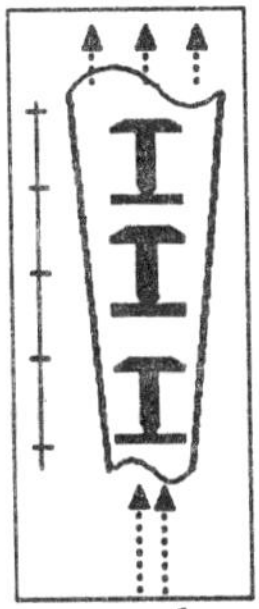

The float rises until the annular area between the float and tube increases sufficiently to allow a state of dynamic equilibrium between the upward differential pressure and buoyancy factors, and downward gravity factors.

The height of the float is an indication of the flow rate. The tube can be calibrated and graduated in appropriate flow units. The rotameter meter typically have a TurnDown Ratio up to 12:1. The accuracy may be as good as 1% of full scale rating. Magnetic floats can be used for alarm and signal transmission functions.

VELOCITY FLOWMETERS

In a velocity flowmeter the flow is calculated by measuring the speed in one or more points in the flow, and integrating the flow speed over the flow area.

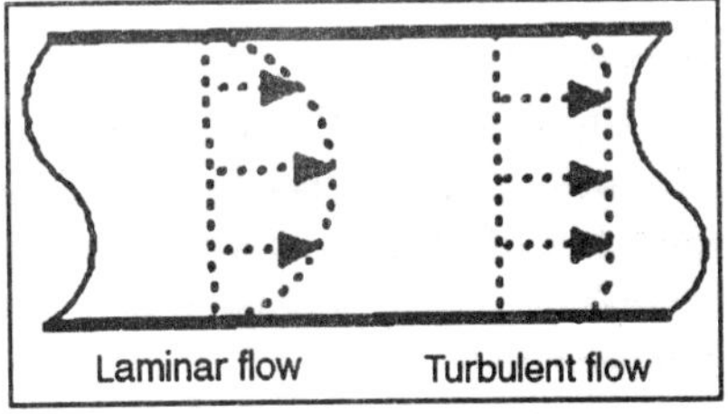

Pitot Tubes

The pitot tube are one the most used (and cheapest) ways to measure fluid flow, especially in air applications as ventilation and HVAC systems, even used in airplanes for the speed measurent.

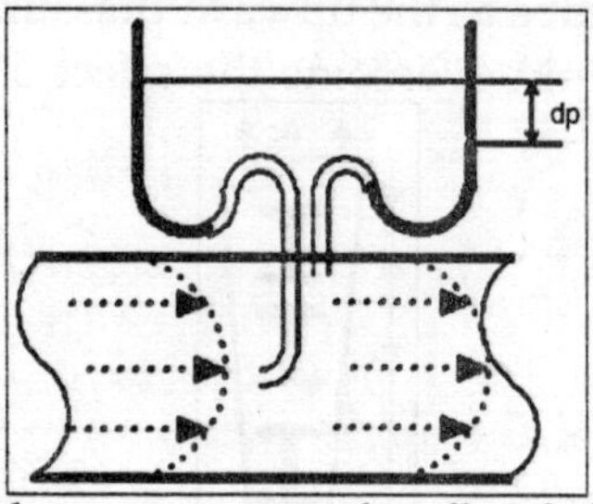

The pitot tube measures the fluid flow velocity by converting the kinetic energy of the flow into potential energy.

The use of the pitot tube is restricted to point measuring. With the "annubar", or multi-orifice pitot probe, the dynamic pressure can be measured across the velocity profile, and the annubar obtains an averaging effect.

Calorimetric Flowmeter

The calorimetric principle for fluid flow measurement is based on two temperature sensors in close contact with the fluid but thermal insulated from each other.

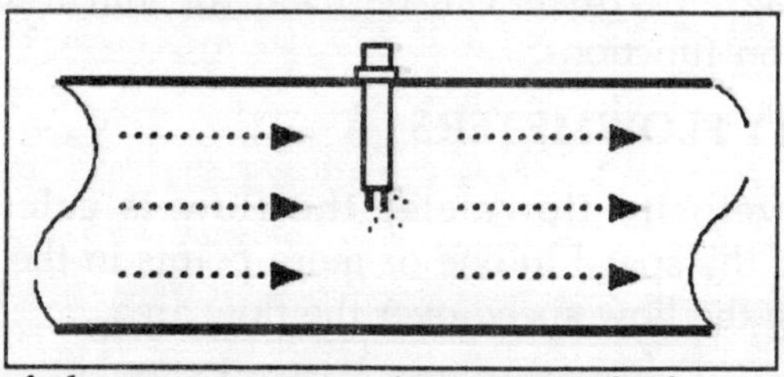

One of the two sensors is constantly heated and the cooling effect of the flowing fluid is used to monitor the flowrate. In a stationary (no flow) fluid condition there is a constant temperature difference between the two temperature sensors. When the fluid flow increases, heat energy is drawn from the heated sensor and the temperature difference between the sensors are reduced. The reduction is proportional to the flow rate of the fluid.

Response times will vary due the thermal conductivity of the fluid. In general lower thermal conductivity require higher velocity for proper measurement. The calorimetric flowmeter can achieve relatively high accuracy at low flow rates.

Turbine Flowmeter

There is many different manufacturing design of turbine flow meters, but in general they are all based on the same simple principle: If a fluid moves through a pipe and acts on the vanes of a turbine, the turbine will start to spin and rotate. The rate of spin is measured to calculate the flow.

The turndown ratios may be more than 100:1 if the turbine meter is calibrated for a single fluid and used at constant conditions. Accuracy may be better than +/-0,1%.

Vortex Flow Meter

An obstruction in a fluid flow creates vortices in a downstream flow. Every obstruction has a critical fluid flow speed at which vortex shedding occurs. Vortex shedding is the instance where alternating low pressure zones are generated in the downstream.

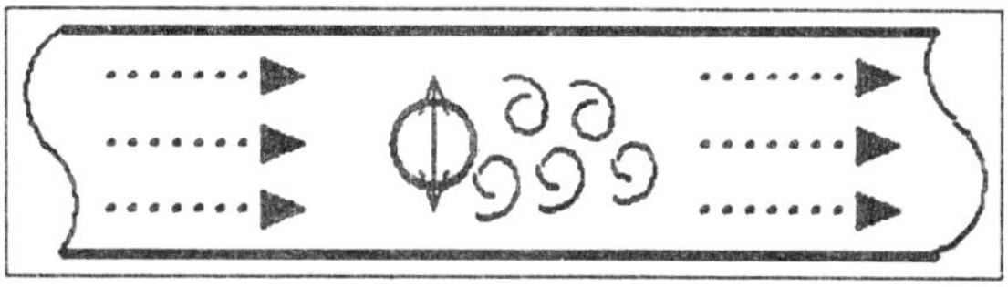

These alternating low pressure zones cause the obstruction to move towards the low pressure zone. With sensors gauging the vortices the strength of the flow can be measured.

- The Vortex Flowmeter Principle - An introduction to the vortex flowmeter principle.

Electromagnetic Flowmeter

An electromagnetic flowmeter operate on Faraday's law of electromagnetic induction that states that a voltage will be induced when a conductor moves through a magnetic field. The liquid serves as the conductor and the magnetic field is created by energized coils outside the flow tube.

The voltage produced is d[illegible] proportional to the flow rate. Two electrodes mounted in the pipe wall detect the voltage which is measured by a secondary element.

Electromagnetic flowmeters can measure difficult and corrosive liquids and slurries, and they can measure flow in both directions with equal accuracy.

Electromagnetic flowmeters have a relatively high power consumption and can only be used for electrical conductive fluids as water.

- The Electromagnetic Flowmeter Principle - An introduction to the electromagnetic flowmeter principle

Ultrasonic Doppler Flowmeter

The effect of motion of a sound source and its effect on the frequency of the sound was observed and described by Christian Johann Doppler.

The frequency of the reflected signal is modified by the velocity and direction of the fluid flow

If a fluid is moving towards a transducer, the frequency of the returning signal will increase. As fluid moves away from a transducer, the frequency of the returning signal decrease.

The frequency difference is equal to the reflected frequency minus the originating frequency and can be use to calculate the fluid flow speed.

- The Ultrasonic Doppler and Time of Flight Flowmeter
- An Ultrasonic Flowmeter Tutorial - A basic tutorial about ultrasonic flowmeters.

Positive Displacement Flowmeter

The positive displacement flowmeter measures process fluid flow by precision-fitted rotors as flow measuring elements. Known and fixed volumes are displaced between the rotors. The rotation of the rotors are proportional to the volume of the fluid being displaced.

The number of rotations of the rotor is counted by an integral electronic pulse transmitter and converted to volume and flow rate.

The positive displacement rotor construction can be done in several ways:

- *Reciprocating piston* meters are of single and multiple-piston types.
- Oval-gear meters have two rotating, oval-shaped gears with synchronized, close fitting teeth. A fixed quantity of liquid passes through the meter for each revolution. Shaft rotation can be monitored to obtain specific flow rates.
- *Nutating disk* meters have moveable disks mounted on a concentric sphere located in spherical side-walled chambers. The pressure of the liquid passing through the measuring chamber causes the disk to rock in a circulating path without rotating about its own axis. It is the only moving part in the measuring chamber.
- *Rotary vane* meters consists of equally divided, rotating impellers, two or more compartments, inside the meter's housings. The impellers are in continuous contact with the casing. A fixed volume of liquid is swept to the meter's outlet from each compartment as the impeller rotates. The revolutions of the impeller are counted and registered in volumetric units.

The positive displacement flowmeter may be used for all relatively nonabrasive fluids such as heating oils, lubrication oils, polymer additives, animal and vegetable fat, printing ink, freon, and many more.

Accuracy may be up to *0.1%* of full rate with a TurnDown of 70:1 or more.

Mass Flowmeters

Mass meters measure the mass flow rate directly.

Thermal Flowmeter

The thermal mass flowmeter operates independent of density, pressure, and viscosity. Thermal meters use a heated sensing element isolated from the fluid flow path where the flow stream conducts heat from the sensing element. The conducted heat is directly proportional to the mass flow rate and the he temperature difference is calculated to mass flow.

The accuracy of the thermal mass flow device depends on the calibrations reliability of the actual process and variations in the temperature, pressure, flow rate, heat capacity and viscosity of the fluid.

Coriolis Flowmeter

Direct mass measurement sets Coriolis flowmeters apart from other technologies. Mass measurement is not sensitive to changes in pressure, temperature, viscosity and density. With the ability to measure liquids, slurries and gases, Coriolis flowmeters are universal meters.

Flowmeter uses the Coriolis effect to measure the amount of mass moving through the element. The fluid to be measured runs through a U-shaped tube that is caused to vibrate in an angular harmonic oscillation.

Due to the Coriolis forces, the tubes will deform and an additional vibration component will be added to the oscillation.

This additional component causes a phase shift on some places of the tubes which can be measured with sensors. The Coriolis flow meters are in general very accurate, better than +/-0,1% with an turndown rate more than 100:1. The Coriolis meter can also be used to measure the fluids density.

Open Channel Flowmeters

A common method of measuring flow through an open channel is to measure the height of the liquid as it passes over an obstruction as a flume or weir in the channel.

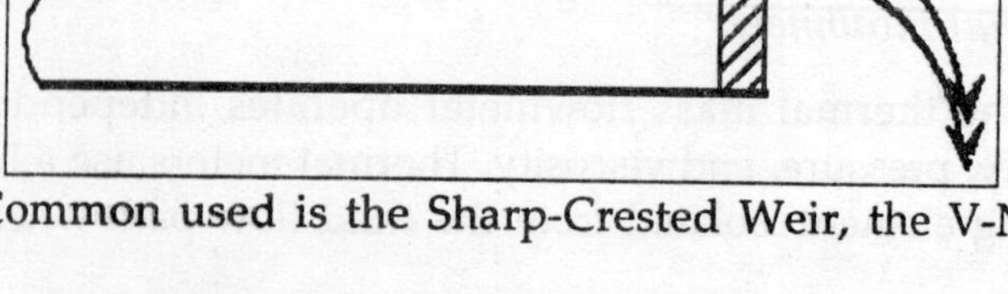

Common used is the Sharp-Crested Weir, the V-Notch

Weir, the Cipolletti weir, the Rectangular-Notch Weir, the Parshall Flume or Venturi Flume.

TYPES OF FLUID MOTION

Fluids, like most other forms of matter, are made up of tiny particles [molecules], which are separated by large spaces.

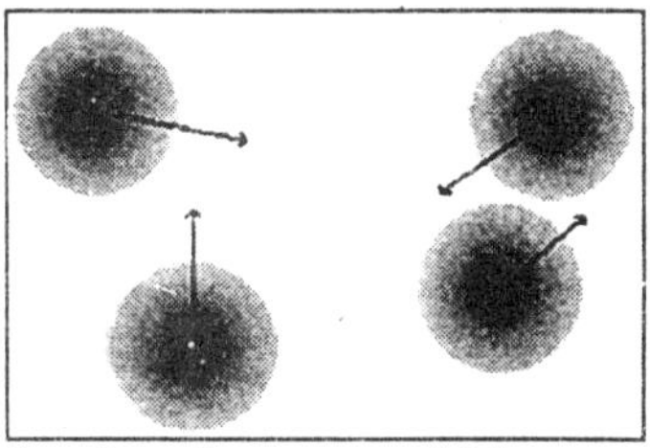

On a really tiny level, we would have to deal with these individual particles. Fortunately, most of the problems we encounter in aerodynamics can be handled by looking at volumes no smaller than, say, (1×10^{-6} m^3), a volume which contains at least a few thousand molecules. [How to calculate: From Chemistry: Avogadro's # gives # of atoms (or molecules) per gram-mole of the gas.

Density of air (mass/volume) can be found from the Perfect Gas Law:

$P = rRT$

where R is gas constant (~ 288 in SI units), and T is temperature, and P is pressure.

⇒ *So we can treat fluid as a "continuum"*: A medium which is pretty uniform in properties at the smallest scales of interest to us.

When one tries to analyse the dynamics of gases at the outer edge of earth's atmosphere, one has to look for methods which don't assume "continuum" fluid mechanics: the distances between molecules are of the same order as the dimensions of interest to us; perhaps meters. The same problem arises at the other end of the size scale: when we try to deal with micro-devices used to sense and control flow behaviour. Here, even at usual sea-level atmospheric conditions, the dimensions of the device may be so small that that they are comparable to the distance between molecules.

Fluid Motion

In the rest of this course we will treat fluids as "continuous media", and forget about those molecules zipping about at random. The smallest "packet" of fluid which we will now consider will have many millions of molecules. The motion of these packets is the net motion of all those molecules zipping around within the packet (think of a school bus moving at 5mph with thirty middle-schoolers conducting a football game inside using each other's backpacks.)

Streamline

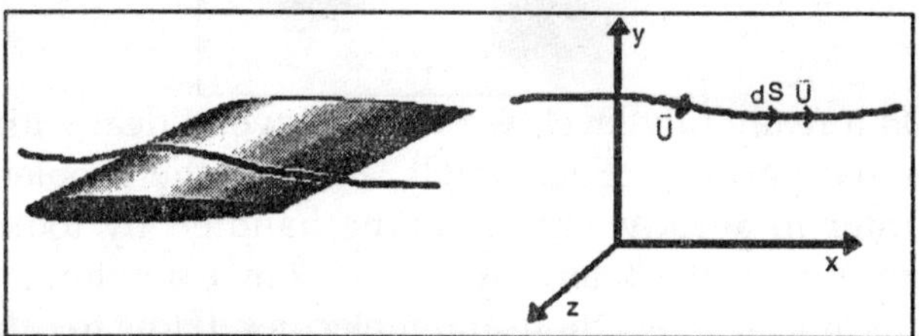

A streamline is a curve whose tangent at any point is in the direction of the velocity vector at that point.

We se that if the flow is not steady, the streamline has little use or significance.

If ds is an elemental vector along the streamline,

$$d\vec{s} = dx\,\vec{i} + dy\,\vec{j} + dz\,\vec{k}$$

$$\vec{U} = u\,\vec{i} + v\,\vec{j} + w\,\vec{k}$$

$$d\vec{s} \times \vec{U} = \begin{vmatrix} i & j & k \\ dx & dy & dz \\ u & v & w \end{vmatrix}$$

$$\Rightarrow i(wdy - vdz) + j(udz - wdx) + k(vdx - udy) = 0$$

$$wdy = vdz \frac{dz}{dy} = \frac{w}{v}$$

$$udz = wdx \frac{dz}{dx} = \frac{w}{u}$$

Equation to a streamline is:

$$vdx = udy \frac{dy}{dx} = \frac{v}{u}$$

As a "packet" of fluid moves along, a combination of 4 things can happen to it: translation, dilatation, rotation and shear.

FLUID MOTION

Thus Basic fluid motion can be described as some combination of

Translation

This is characterized by the velocity

$$\vec{U} = u\vec{i} + v\vec{j} + w\vec{k}$$

$$\rho = \frac{P}{RT}$$

Dilatation

[volume change]

$\vec{u}, \rho, p, e, h$

Rotation

[About one, two or 3 axes].

$$\omega = \frac{1}{2}\left[\left(\frac{dw}{dy} - \frac{dv}{dz}\right)\vec{i} + \left(\frac{du}{dz} - \frac{dw}{dx}\right)j + \left(\frac{dv}{dx} - \frac{du}{dy}\right)k\right]$$

Note that rotation is a vector. We usually use a quantity called "vorticity" which is twice the rotation vector, to describe the amount of rotation in flows.

Shear Strain

The quantity "strain" thus has nine components.

They are:

- e_{xx}, e_{yy}, e_{zz} These are "normal strain" components.
- e_{xy}, e_{yz}, e_{zx} These are "shear strain" components.
- e_{yx}, e_{zy}, e_{xz} These are equal to the corresponding "shear strain" components above.

If this discussion brings back memories of your solid mechanics / statics course, there is good reason for it. The difference comes in the relation between strain and the stress:

in the case of solids, the stress depends on the strain. In the case of fluids, the stress depends on the rate of strain.

Circulation

Defined as $d\vec{s}$ integrated around a closed contour. The negative sign is included such that positive circulation on a body corresponds to positive lift, and the integral is evaluated counter-clockwise.

From the preceding discussion, we see that r will be zero unless there is some vorticity, $\vec{\omega}$ contained within the contour. *r is an extremely useful quantity*: it helps us calculate lift, vortex strength, etc.

Important Points

$$d\vec{s}\times\vec{U}=0$$

The circulation around a closed contour with net rotation and/or shear will be non-zero. It is, however, always possible to have a combination of rotation and/or shear that gives a zerocirculation.

CLASSIFICATION OF FLUID FLOW

Fluid flow can be very smooth, calm, and regular, but generally the flow of a fluid is not so disciplined- it becomes a vagabond and starts flowing in random patterns. What are different flow patterns, how are flows classified, and what are the fluid paths in those patterns?

While studying the motion of a rigid body we do not have to bother about the relative motion of the particles of the rigid body as they are very firmly fixed to each other and move as a whole.

But for the study of the motion of fluids, things are not so simple because the fluid particles are attached with each other with very weak forces. There are various relative motions and a lot of possibilities for relative motion between the fluid particles.

To make things somewhat simple or for making the flow analysis feasible, fluid flow is visualized as a composition of fluid elements. These elements are defined by using certain similarities or patterns and mathematics is applied to them to study fluid flow comprehensively.

Rotational or Irrotational Flow

To classify any flow as Rotational or Irrotational the angular motion of the fluid elements is analysed. If the angle between the two intersecting lines of the boundary of the fluid element changes while moving in the flow, then the flow is a Rotational Flow. But if the fluid element rotates as a whole and there is no change in angles between the boundary lines then the flow cannot be Rotational Flow, so it is Irrotational Flow.

This means that there should be some deformation in the fluid element in a Rotational Flow. Such deformation of the fluid element or the shear strain is necessarily caused by tangential forces or shear stresses. Shear stresses are caused by viscosity, thus the flow of viscous fluids is rotational. But this does not mean that the flow of non-viscous or ideal fluid is always irrotational. The flow of ideal fluids can be rotational by external work or heat interaction.

Laminar or Turbulent Flow

Laminar Flow

The flow of a fluid moving with a moderate speed has fluid layers moving past other layers as if some sheets are moving over other layers. Such flow of fluids is called Laminar Flow. In Laminar Flow viscous shear stresses act between these layers of the fluid which defines the velociity distribution among these layers of flow. In Laminar Flows the shear stresses are defined by Newton's equation for shear stress.

Turbulent Flow

As the flow speed of the otherwise calm layers increases, these smoothly moving layers start moving randomly, and with further increase in flow velocity, the flow of fluid particles becomes completely random and no such laminar layers exist any more. Shear stresses in the Turbulent Flow are more than those in Laminar Flow.

A dimensionless parameter, Reynolds Number, is defined as the ratio of inertial and viscous force to characterize these two types of flow patterns. With increase in flow velocity the intial forces increase so the Reynonlds Number. For moderate flcws the Reynolds Number is below 2000 and for Turbulent Flows it is well above 2300. For the transition region between the two types the Reynolds Number varies between 2000-4000.

Fundamentals of Fluid Flow

How does a flowing fluid behaves,and how do the different parameters vary in a flowing fluid? Fluid flow analysis involves theoretical and experimental study of the flow of fluid by defining flow models and observation of variation of parameters. This series will discuss Fundamentals of Fluid Flow.

CONTINUUM AND MOLECULAR

SHOCK WAVES AND CONTINUUM FLOW MODELS

Since the conservation equation describes flow and density as a function of distance and time, one can immediately see that continuum modeling is superior to input-output models used in practice (which are only one dimensional, because they essentially ignore space). In addition, because flow is assumed to be a function of density, continuum models have a second major advantage, (e.g. compressibility).

The simple continuum model referred to in this text consists of the conservation equation and the equation of state (speed-density or flow density relationship). If these equations are solved together with the basic traffic flow equation (flow equals density times speed), then we can obtain speed, flow, density at any time and point of the roadway.

Knowing these basic traffic flow variables we know the state of the traffic system and can derive measures of effectiveness, such as delays stops, total travel, travel time, and others that allow engineers to evaluate how well the system is performing....

A shock wave is a discontinuity of flow or density, and

has the physical implication that cars change speeds abruptly without time to accelerate or decelerate. This is an unnatural behaviour that could be eliminated by considering high order continuum models.

These models add a momentum equation that accounts for the acceleration and inertia characteristics of traffic mass. In this manner, shock waves are smoothed out and the equilibrium assumption is removed.... In spite of this improvement, the most widely known high order models still require an equilibrium speed-density relationship....

$$\left[uf - 2uf\frac{k}{kj}\right]\frac{\partial k}{\partial x} + \frac{\partial k}{\partial t} = 0$$

where,

u_f represents the free flow speed

k_j the jam density is a first order quasi-linear, partial differential equation which can be solved by the method of characteristics.

In practical terms, the solution suggests that:

- The density k is a constant along a family of curves called characteristics or waves; a wave represents the motion (propagation) of a change in flow and density along the roadway.
- The characteristics are straight lines emanating from the boundaries of the time-space domain.
- *The slope of the characteristics is:*

$$\frac{dx}{dt} = f(k) + k[f(k)] = \frac{dq}{dk}$$

- This implies that the characteristics have slope equal to the tangent of the flow-density curve at the point representing the flow conditions at the boundary from which the characteristic emanates.
- The density at any point x,t of the time space domain is found by drawing the proper characteristic passing through that point.
- The characteristics carry the value of density (and flow) at the boundary from which they emanate.
- When two characteristic lines intersect, then density

at this point should have two values which is physically unrealizable; t...s discrepancy is explained by the generation of shock waves. In short, when two characteristics intersect, a shock wave is generated and the characteristics terminate. A shock then represents a mathematical discontinuity (abrupt change) in *k*, *q*, or*u*.

The speed of the shock wave is:

$$u_w = \frac{q_d - q_u}{k_d - k_u}$$

where k_d, q_d represent downstream and k_u, q_u upstream flow conditions. In the flow concentration curve, the shock wave speed is represented by the slope of the line connecting the two flow conditions (i.e., upstream and downstream).

It should be noted that when u_w is positive, the shock wave moves downstream with respect to the roadway; conversely, when u_w is negative, the shock is moving upstream.

Furthermore, the mere fact that a difference exists in flow conditions upstream and downstream of a point does not imply that a shock wave is present unless the characteristics intersect.

Generally this occurs only when the downstream density is higher than upstream. When density downstream is lower than upstream, we have diffusion of flow similar to that observed when a queue is discharging. When downstream density is higher than upstream, then shock waves are generated and queues are generally being built even though they might be moving downstream.

Figure demonstrates the use of traffic waves in identifying the occurrence of a shock wave and following its trajectory. The process follows the steps of the solution of the conservation equation as outlined above. The top of the figure represents a glow-concentration curve; the bottom figure represents trajectories of the traffic waves. On the *q-k* curve, point A represents a situation where traffic flows at near capacity implying that speed is well below the free-flow speed.

Point B represents an uncongested condition where traffic flows at a higher speed because of the lower density. Tangents at points A and B represent the wave velocities of these two situations. The areas where conditions A and B prevail are shown by the characteristics drawn in the bottom. This figure assumes that the faster flow of point B occurs later in time than that of point A; therefore, the characteristics (waves) of point B will eventually intersect with those of point A.

The intersection of these two sets of waves has a slope equal to the chord connecting the two points on the *q-k* curve, and this intersection represents the path of the shock wave shown at the bottom.

It is necessary to clarify that the waves of the time-space are not the trajectories of vehicles but lines of constant flow and speed showing the propagation of conditions A and B. The velocities of individual vehicles within A and B are higher because the speed of the traffic stream is represented by the line connecting the origin with A and B in the *q-k* curve.

FREE MOLECULAR FLOW

Definition

A flow regime in aerodynamics in which molecules emitted from an object, as it passes through a resistive medium, do not affect the flow of oncoming molecules by scattering interactions, i.e., the mean free path of the emitted molecules in much longer than a characteristic linear dimension of an object.2. Flow about a body in which the number of collisions between the molecules of the fluid is negligible compared with the collisions between these molecules and the body. Also called free molecular flow. See rarefied gas dynamics

The transmission probabilities and the number of molecules reflecting from the wall were computed using a Monte Carlo method for free molecule flow through a conduit containing an orifice of zero thickness. The computed transmission probabilities were compared with a composite transmission rule and a series-resistance equation.

The distribution of the molecular number density on the conduit wall along the conduit axis was determined by assuming that the ratio of the molecular number flux reflecting from the conduit wall to that entering into the conduit inlet corresponds to the ratio of the molecular number densities at each position.

When the orifice was placed halfway down the length of the conduit, the composite rule and the series-resistance equation, respectively, agreed well with the computed transmission probabilities for the conduits with the aspect ratio $L/R \geq 10$, and the cross-sectional area ratio of tube and orifice opening $\beta \geq 10$, and for the conduit with $\beta \geq 40$.

The molecular number density distributions were symmetric with respect to the midpoint, and had a minimum in the upstream side and a maximum in the downstream side. The molecular number density distributions exhibited jumps across the orifice, as well as at the inlet and exit of the conduit.

Gas-flow phenomenon at low pressures or in small channels when the mean free path is of the same order of magnitude as the channel diameter; a gas molecule thus migrates along the channel independent of other gas molecules present.

UNIFORM AND NON-UNIFORM

There are three categories or generic ways in which motion, in the strict sense, can occur. For whatever is moved, is changed either in its place, or in its quantity, or in its quality. And since, in general, any successive motion whatever is fast or slow, and since no single method of determining velocity is applicable in the same sense to all three kinds of motion, it will be suitable to show how any change of this sort may be distinguished from another change of its own kind, with respect to speed or slowness. And because local motion is prior in nature to the other kinds, as the primary kind, we will carry out our intention in this section, with respect to local motion, before treating of the other kinds.

MEASURE OF UNIFORM VELOCITY

Although change of place is of diverse kinds, and is varied according to several essential as well as accidental differences, yet it will suffice for our purposes to distinguish uniform motion from nonuniform motion. Of local motions, then, that motion is called uniform in which an equal distance is continuously traversed with equal velocity in an equal part of time. Nonuniform motion can, on the other hand, be varied in an infinite number of ways, both with respect to the magnitude, and with respect to the time.

In uniform motion, then, the velocity of a magnitude as a whole is in all cases measured (*metietur*) by the linear path traversed by the point which is in most rapid motion, if there is such a point. And according as the position of this point is changed uniformly or non uniformly, the complete motion of the whole body is said to be uniform or difform (non uniform). Thus, given a magnitude whose most rapidly moving point is moved uniformly, then, however much the remaining points may be moving nonuniformly, that magnitude as a whole is said to be in uniform movement...

MEASURE OF NON UNIFORM VELOCITY

In non uniform motion, however, the velocity at any given instant will be measured (*attendetur*) by the path which *would* be described by the most rapidly moving point if, in a period of time, it were moved uniformly at the same degree of velocity (*uniformiter illo gradu velocitatis*) with which it is moved in that given instant, whatever [instant] be assigned. For suppose that the point A will be continuously accelerated throughout an hour.

It is not then necessary that, in any instant of that hour as a whole, its velocity be measured by the line which that point describes in that hour. For it is not required, in order that any two points or any other two moving things be moved at equal velocity, that they should traverse equal spaces in an equal time; but it is possible that they traverse unequal spaces, in whatever proportion you may please. For suppose that point A is moved continuously and uniformly at C degrees of velocity, for an hour, and that it traverses a distance of a foot.

And suppose that point B commences to move, from rest and in the first half of that hour *accelerates* its velocity to C degrees, while in the second half hour it decelerates from this velocity to rest. It is then found that at the middle instant of the whole hour point B will be moving at C degrees of velocity, and will fully equal the velocity of the point A. And yet, at the middle instant of that hour, B will not have traversed as long a line as A, other things being equal.

In similar manner, the point B, traversing a finite line as small as you please, can be accelerated in its motion beyond any limit; for, in the first proportional part of that time, it may have a certain velocity, and in the second proportional part, twice that velocity, and in the third proportional part, four times that velocity, and so on without limit.

From this it clearly follows, that such a nonuniform or instantaneous velocity (*velocitas instantanea*) is not measured by the distance traversed, but by the distance which would be traversed by such a point, *if* it were moved uniformly over such or such a period of time at that degree of velocity with which it is moved in that assigned instant.

MEASURE OF UNIFORM ACCELERATION

With regard to the acceleration (*intensio*) and deceleration (*remisso*) of local motion, however, it is to be noted that there are two ways in which a motion may be accelerated or decelerated: namely, uniformly, or nonuniformly.

For any motion whatever is *uniformly accelerated* (*uniformiter intenditur*) if, in each of any equal parts of the time whatsoever, it acquires an equal increment (*latitudo*) of velocity.

But a motion is *nonuniformly accelerated or decelerated,* when it acquires or loses a greater increment of velocity in one part of the time than in another equal part.

In view of this, it is sufficiently apparent that when the latitude of motion or velocity is infinite, it is impossible for any body to acquire that latitude uniformly, in any finite time. And since any degree of velocity whatsoever differs by a finite amount from zero velocity, or from the privative limit of the intensive scale, which is rest – therefore any mobile body may be uniformly accelerated from rest to any assigned degree of velocity; and likewise, it may be decelerated uniformly from any assigned velocity, to rest. And, in general, both kinds of change may take place uniformly, from any degree of velocity to any other degree.

In this connection, it should be noted that just as there is no degree of velocity by which, with continuously uniform motion, a greater distance is traversed in one part of the time than in another equal part of the time, so there is no latitude (i.e., increment, *latitudo*) of velocity between zero degree [of velocity] and some finite degree, through which a greater distance is traversed by uniformly accelerated motion in some given time, than would be traversed in an equal time by a uniformly decelerated motion of that latitude.

For whether it commences from zero degree or from some [finite] degree, every latitude, as long as it is terminated at some finite degree, and as long as it is acquired or lost uniformly, will correspond to its mean degree [of velocity]. Thus the moving body, acquiring or losing this latitude uniformly during some assigned period of time, will traverse a distance exactly equal to what it would traverse in an equal period of time if it were moved uniformly at its mean degree [of velocity].

- For of every such latitude commencing from rest and terminating at some [finite] degree [of velocity], the mean degree is one-half the terminal degree [of velocity] of that same latitude.
- From this it follows that the mean degree of any latitude bounded by two degrees (taken either inclusively or exclusively) is more than half the more intense degree bounding that latitude.
- From the foregoing it follows that when any mobile body is uniformly accelerated from rest to some given degree [of velocity], it will in that time traverse one-half the distance that it would traverse if, in that same time, it were moved uniformly at the degree [of velocity] terminating that latitude. For that motion, as a whole, will correspond to the mean degree of that latitude, which is precisely one-half that degree which is its terminal velocity.
- It also follows in the same way that when any

moving body is uniformly accelerated from some degree [of velocity] (taken exclusively) to another degree inclusively or exclusively, it will traverse more than one-half the distance which it would traverse with a uniform motion, in an equal time, at the degree [of velocity] at which it arrives in the accelerated motion. For that whole motion will correspond to its mean degree [of velocity], which is greater than one-half of the degree [of velocity] terminating the latitude to be acquired; for although a nonuniform motion of this kind will likewise correspond to its mean degree [of velocity] contained in this latitude being acquired, and, likewise, it will be as slow.

- To prove, however, that in the case of acceleration from rest to a finite degree [of velocity], the mean degree [of velocity] is exactly one-half the terminal degree [of velocity], it should be known that if any three terms are in continuous proportion, the ratio of the first to the second, or of the second to the third, will be the same as the ratio of the difference between the first and the middle, to the difference between the middle and the third; as when the terms are 4, 2, 1; 9, 3, 1; 9, 6, 4. For as 4 is to 2, or as 2 is to 1, so is the proportion of the difference between 4 and 2 to the difference between 2 and 1, because the difference between 4 and 2 is 2, while that between 2 and 1 is 1; and so with the other cases.

Let there be assigned, then, some term under which there is an infinite series of other terms which are in continuous proportion according to the ratio 2 to 1. Let each term be considered in relation to the one immediately following it. Then, whatever is the difference between the first term assigned and the second, such precisely will be the sum of all the differences between the succeeding terms. For whatever is the amount of the first proportional part of any continuum or of any finite quantity, such precisely is the amount of the sum of all the remaining proportional parts of it.

Since, therefore, every latitude is a certain quantity, and since, in general, in every quantity the mean is equidistant from the extremes, so the mean degree of any finite latitude whatsoever is equidistant from the two extremes, whether these two extremes be both of them positive degrees, or one of them be a certain degree and the other a privation of it or zero degree.

But, as has already been shown, given some degree under which there is an infinite series of other degrees in continuous proportion, and letting each term be considered in relation to the one next to it, then the difference or latitude between the first and the second degree – the one, namely, that is half the first – will be equal to the latitude composed of all the differences or latitudes between all the remaining degrees - namely those which come after the first two. Hence, exactly equally and by an equal latitude that second degree, which is related to the first as a half to its double, will differ from that double as that same degree differs from zero degree or from the opposite extreme of the given magnitude.

And so it is proved universally for every latitude commencing from zero degree and terminating at some finite degree, and containing some degree and half that degree and one-quarter of that degree, and so on to infinity, that its mean degree is exactly one-half its terminal degree. Hence this is not only true of the latitude of velocity of motion commencing from zero degree [of velocity], but it could be proved and argued in just the same way in the case of latitudes of heat, cold, light, and other such qualities.

- With respect, however, to the distance traversed in a uniformly accelerated motion commencing from zero degree [of velocity] and terminating at some finite degree [of velocity], it has already been said that the motion as a whole, or its whole acquisition, will correspond to its mean degree [of velocity]. The same thing holds true if the latitude of motion is uniformly acquired from some degree [of velocity] in an exclusive sense, and is terminated at some finite degree [of velocity].

"From the foregoing it can be sufficiently determined for this kind of uniform acceleration or deceleration how great a distance will be traversed, other things being equal, in the first half of the time and how much in the second half. For when the acceleration of a motion takes place uniformly from zero degree [of velocity] to some degree [of velocity], the distance it will traverse in the first half of the time will be exactly one-third of that which it will traverse in the second half of the time.

And if, contrariwise, from that same degree [of velocity] or from any other degree whatsoever, there is uniform deceleration to zero degree [of velocity], exactly three times the distance will be traversed in the first half of the time, as will be traversed in the second half. For every motion as a whole, completed in a whole period of time, corresponds to its mean degree [of velocity] – namely, to the degree it will have at the middle instant of the time. And the second half of the motion in question will correspond to the mean degree of the second half of that same motion, which is one-fourth of the degree [of velocity] terminating that latitude.

Consequently, since this second half will last only through half the time, exactly one-fourth of the distance will be traversed in that second half as will be traversed in the whole motion. Therefore, of the whole distance being traversed by the whole motion, three-quarters will be traversed in the first half of the whole motion, and the last quarter will be traversed in its second half. It follows, consequently, that in this type of uniform intension and remission of a motion from some degree [of velocity] to zero degree, or from zero degree to some degree, exactly three times as much distance is traversed in the more intense half of the latitude as in the less intense half.

- But any motion can be uniformly accelerated or decelerated from some degree [of velocity] to another degree in an endless number of ways, because it may be from some degree to a degree half of that, or to a degree one-fourth of it, or one-fifth, or to a degree two-thirds of that degree, or three quarters of it, and so on. Consequently there can be no universal numerical value by which one will be able to determine, for all cases, how much more distance would be traversed in the first half of this sort of acceleration or deceleration than in the second half, because, according to the diversity of the extreme degrees [of velocity], there will be diverse proportions of distance traversed in the first half of the time to distance traversed in the second half.

But if the extreme degrees [of velocity] are determined, so that it is known, for instance, that so much distance would be traversed in such or such a time by a uniform motion at the more intense limiting degree [of velocity], and if this is likewise known with respect to the less intense limiting degree [of velocity], then it will be known by calculation how much would be traversed in the first half and also how much in the second. For, if the extreme degrees [of velocity] are known in this way, the mean degree [of velocity] of these can be obtained, and also the mean degree between that mean degree and the more intense degree terminating the latitude. But a calculation of this kind offers more difficulty than advantage.

And it is sufficient, therefore, for every case of this kind, to state as a general law, that more distance will be traversed by the more intense half of such a latitude than by the less intense half – as much more, namely, as would be [the excess of distance] traversed by the mean degree [of velocity] of this more intense half, if it moved in a time equal to that in which this half is acquired or lost uniformly, over that [distance which] would be traversed by the mean degree [of velocity] of the less intense half, in the same time.

- But as concerns nonuniform acceleration or deceleration, whether from some degree [of velocity] to zero degree or *vice versa,* or from one degree to some other degree, there can be no rule determining the distance traversed in such or such time, or determining the intrinsic degree to which such a latitude of motion, acquired or lost nonuniformly, will correspond. For just as such a nonuniform acceleration or deceleration could vary in an infinite number of ways, so also that motion as a whole could correspond to an infinite number of intrinsic degrees [of velocity] of its latitude – indeed, to any intrinsic degree whatsoever, of the latitude thus acquired or lost.

In general, therefore, the degree [of velocity] terminating such a latitude at its more intense limit is the most remiss degree [of velocity], beyond the other limit (i.e., the most remiss extreme) of the latitude, to which such a nonuniformly nonuniform motion as a whole *cannot* correspond; and the degree [of velocity] terminating that latitude at its more remiss limit is the most intense degree [of velocity] beneath the upper limit of the same latitude, to which such a nonuniformly nonuniform motion *cannot* correspond. Consequently, it is not possible for such a motion as a whole to correspond to such a remiss degree (as that of the lower limit); nor to such an intense degree (as the upper limit).

NON-UNIFORM ACCELERATION

Students will have already considered uniformly accelerated motion. This demonstration (or experiment) uses a similar method to consider non-uniform motion. You can conclude the demonstration by discussing the relationships shown on the graphs, stressing that these hold for both uniform and non-uniform motion.

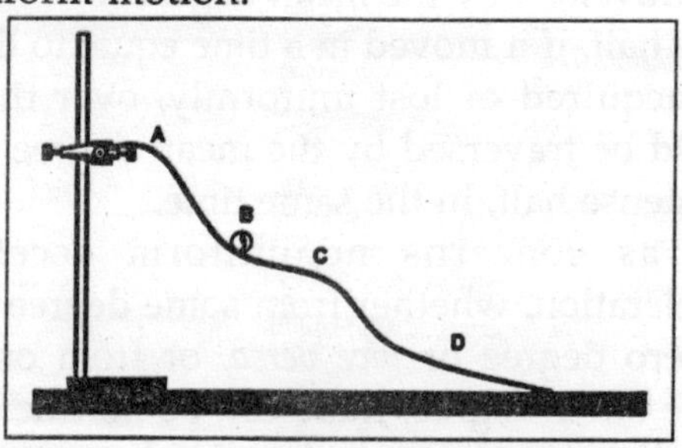

DEVELOPING EQUATIONS OF MOTION

Mathematicians will enjoy the mild challenge of developing SUVAT equations whereas weaker or non-mathematical students may find the activity surprisingly difficult. It is therefore best to proceed through the activity at a reasonable pace so that you concentrate on the results and using the SUVAT equations.

UNIFORM AND NON UNIFORM CIRCULAR MOTION

Angular Velocity And Angular Acceleration

The angular velocity (or angular frequency) of a body in a uniform circular motion is the angle swept out by the radius vector per second. If the radius vector sweeps out an angle $\Delta\theta$ (measured in radians) in a time interval Δt, the angular velocity is given by

$$\omega = \frac{\lim}{At \rightarrow \theta}\left(\frac{\Delta\theta}{\Delta t}\right) = \frac{d\theta}{dt}$$

where ω is expressed in rad s^{-1}. Linear velocity (v), angular velocity (ω) and the radius (r) of the circular motion are related as $v = r\omega$

Radian measure of an angle:

If the arc of a circle equals the radius, then the angle subtended by that arc at the centre of the circle is called a radian; it is equal to about 57.3^o (see Fig.)

Angular Acceleration

The angular acceleration á is given by,$\alpha = d\omega/dt$

Linear acceleration (a), angular acceleration (α) and radius (r) of the circular motion are related as $a = r\alpha$ Angular acceleration is expressed in rad s^{-2}.

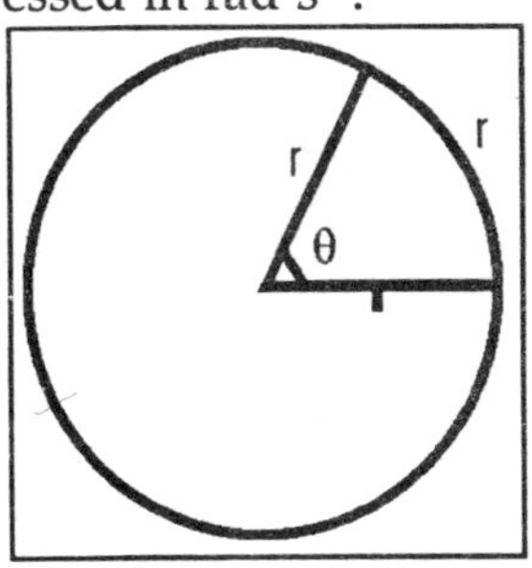

Centripetal Acceleration

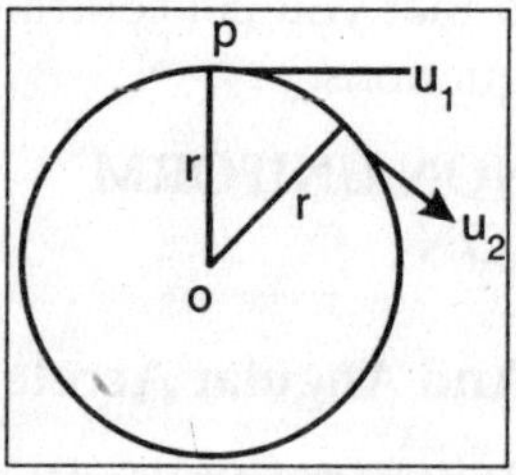

If a body moves in a circle at a constant speed, it is said to be in uniform circular motion. In such a motion, the magnitude of the velocity (i.e. speed) is constant but the direction of the velocity vector is continually changing. Thus the velocity is changing with time.

Hence the motion of the body is accelerated. The acceleration is directed towards the centre of the circle and is called centripetal acceleration. The magnitude of the centripetal acceleration is given by $a_c = \omega v$

where

ω is the angular velocity (or angular frequency)

v is the speed along the circle.

Since $v = r\omega$,

we have,

$a_c = wv = w^2 r = v^2/r$

where

r is the radius of the circular path.

The angular frequency is related to time period T and frequency v as, $\omega = 2\pi/T = 2\pi v$

Therefore, centripetal acceleration is also given by (since $v = 2\pi r/T$),

$a_c = \omega v = 4\pi^2 r/T^2 = 4\pi^2 r v^2$

UNIFORM CIRCULAR MOTION

- Velocity remains constant in magnitude but varies in direction.
- The acceleration is always normal to the velocity vector.
- The acceleration is always directed towards the centre of the circular path.

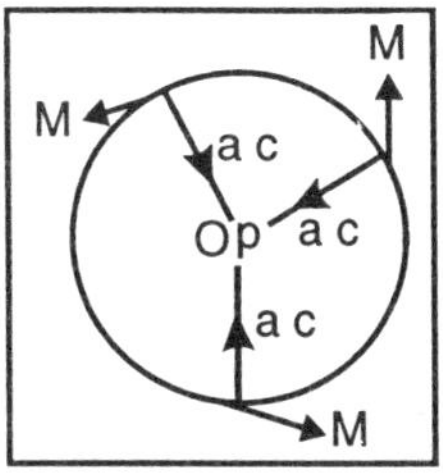

Comparison of uniform circular motion with straight line motion and projectile motion:

NON-UNIFORM CIRCULAR MOTION

- The velocity changes both in magnitude as well as in direction
- The velocity vector is always tangential to the path
- The acceleration vector is not perpendicular to the velocity vector
- The acceleration vector has two components
  - Tangential acceleration a_t changes the magnitude of velocity vector, i.e. $a_t = dv/dt$
  - Normal acceleration or centripetal acceleration a_c changes the direction of the velocity vector, i.e. $a_c = v^2/r$
  - The total acceleration is the vector sum of the tangential and centripetal acceleration $a = \Phi at^2 + ac^2$

LAMINAR AND TURBULENT FLOW

The nature of the flow of a fluid by injecting a fine filament of dye into the stream of flow and taking note of what happens to this filament.

It was found in experiments that at low velocities the dye filament remained intact and that the filaments made parallel lines in the stream of flow. This is known as Laminar flow (or viscous or streamline).

If the velocity of flow is gradually increased, the dye filament is eventually broken up and spread over the cross section of the pipe. This is turbulent flow, in which the particles of fluid are not moving in parallel lines but are moving across the general direction of flow.

If a fluid particle in a stream is disturbed, its inertia will tend to move it in a new direction, however the viscous forces from the surrounding fluid will tend to move it in the general direction of flow. If the shear forces are large enough to overcome any deviation, then we have viscous or laminar flow. However, if the shear forces are relatively weaker, and not sufficient to overcome the inertia of the particles, then we have turbulent flow.

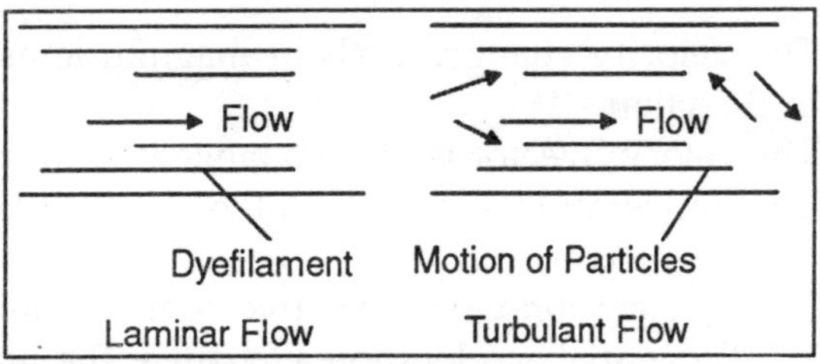

Therefore it is the ratio of the inertia to the viscous forces which determines whether flow will be laminar or turbulent.

The ratio of the inertia forces to the viscour forces is given by:

$\rho \frac{cl}{\mu}$ (Reynolds Number)

Therefore, it is the Reynolds number which determines whether a flow will be laminar or turbulent.

As Kinematic Viscosity $(\nu) = \frac{\mu}{\rho}$,

Reynolds number can also be expressed as c l

νl is the characteristic dimension in the system e.g. the diameter of the pipe in which the fluid is flowing.

Experiments have shown that if the Reynolds number is less than 2000, then the flow will be laminar. If it is above 2000 it is likely to be turbulent, however turbulent flow can exist at much higher Reynolds numbers.

Example: Water flows through a pipe 30mm diameter at a velocity of 10 m/s. If the dynamic viscosity of water is taken as 1.30×10^{-3} kg/ms and its density as 1000 kg/m^3, determine whether the flow will be laminar or turbulent.

Reynolds number,

$$= \rho \frac{c1}{\mu}$$

$$= \frac{1000 \times 10 \times 0.03}{1.3 \times 10^{-3}}$$

$$= 230769$$

This is significantly larger than 2000 and therefore the flow will be turbulent.

Problem If oil of specific gravity 0.85 and dynamic viscosity 14×10^{-2} kg/ms is pumped through the pipe at the same viscocity, what type of flow will occur.

If we were to take a pipe of free flowing water and inject a dye into the middle of the stream, what would we expect to happen?

This,

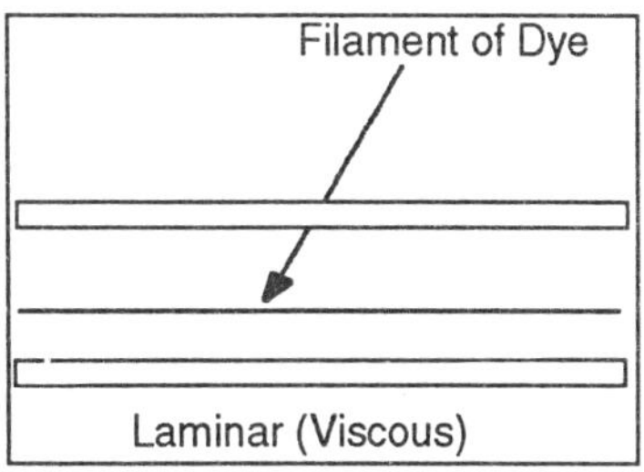

this

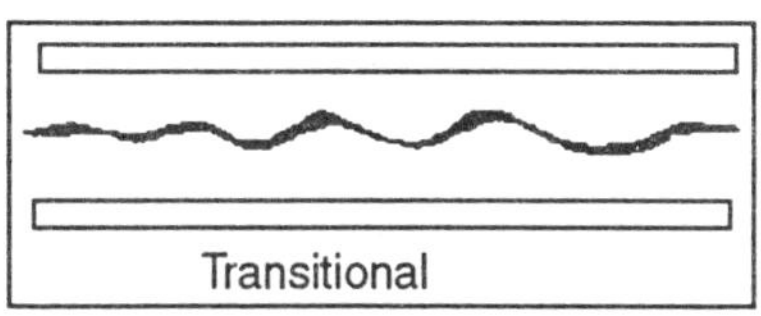

or this

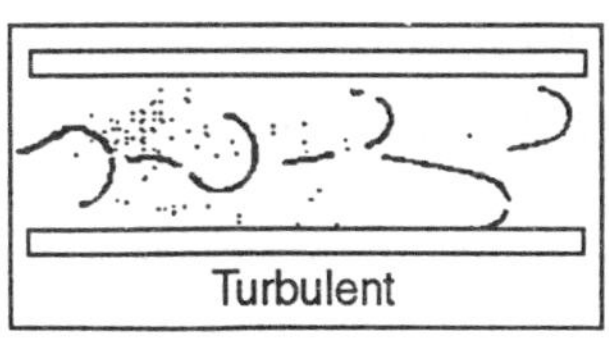

Actually both would happen - but for different flow rates. The top occurs when the fluid is flowing fast and the lower when it is flowing slowly.

The top situation is known as turbulent flow and the lower as laminar flow.

In laminar flow the motion of the particles of fluid is very orderly with all particles moving in straight lines parallel to the pipe walls.But what is fast or slow? And at what speed does the flow pattern change? And why might we want to know this?

The phenomenon was first investigated in the 1880s by Osbourne Reynolds in an experiment which has become a classic in fluid mechanics.

$$\frac{\rho u d}{\mu}$$

He used a tank arranged as above with a pipe taking water from the centre into which he injected a dye through a needle. After many experiments he saw that this expression

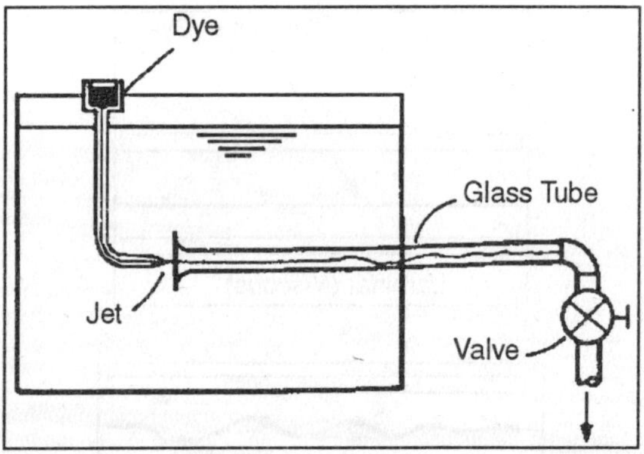

where ρ = density, u = mean velocity, d = diameter and μ = viscosity

would help predict the change in flow type. If the value is less than about 2000 then flow is laminar, if greater than 4000 then turbulent and in between these then in the transition zone.

This value is known as the Reynolds number, Re:

$$Re = \frac{\rho u d}{\mu}$$

Laminar flow: Re < 2000

Transitional flow: 2000 < Re < 4000
Turbulent flow: Re > 4000

Units of Reynolds number

We can fill in the equation with SI units:

$\rho = kg/m^3, u = m/s, d = m$

$\mu = Ns/M^2 = kg/ms$

$$Re = \frac{\rho ud}{\mu} = \frac{kg}{m^3}\frac{m}{s}\frac{m}{1}\frac{m}{kg} = 1$$

i.e. it has no units. A quantity that has no units is known as a non-dimensional (or dimensionless) quantity. Thus the Reynolds number, Re, is a non-dimensional number.

We can go through an example to discover at what velocity the flow in a pipe stops being laminar.

If the pipe and the fluid have the following properties:
water density $\rho = 1000$ kg/m^3
pipe diameter $d = 0.5$m
(dynamic) viscosity, $\mu = 0.55x10^3$ Ns/m^2

We want to know the maximum velocity when the Re is 2000.

$$Re = \frac{\rho ud}{\mu} = 2000$$

$$\mu = \frac{2000\mu}{\rho d} = \frac{2000 \times 0.55 \times 10^{-3}}{1000 \times 0.5}$$

$$m = 0.0022 \text{ m/s}$$

If this were a pipe in a house central heating system, where the pipe diameter is typically 0.015m, the limiting velocity for laminar flow would be, 0.0733 m/s.

Both of these are very slow. In practice it very rarely occurs in a piped water system - the velocities of flow are much greater. Laminar flow does occur in situations with fluids of greater viscosity - e.g. in bearing with oil as the lubricant.

At small values of Re above 2000 the flow exhibits small instabilities. At values of about 4000 we can say that the flow is truly turbulent. Over the past 100 years since this experiment, numerous more experiments have shown this phenomenon of limits of Re for many different Newtonian fluids - including gasses.

Abstract Number Mean

We can say that the number has a physical meaning, by doing so it helps to understand some of the reasons for the changes from laminar to turbulent flow.

$$Re = \frac{\rho u d}{\mu}$$

$$= \frac{\text{inertial forces}}{\text{viscous forces}}$$

It can be interpreted that when the inertial forces dominate over the viscous forces (when the fluid is flowing faster and Re is larger) then the flow is turbulent. When the viscous forces are dominant (slow flow, low Re) they are sufficient enough to keep all the fluid particles in line, then the flow is laminar.

LAMINAR FLOW

Laminar flow is also referred to as streamline or viscous flow. These terms are descriptive ofthe flow because, in laminar flow:

(1) Layers of water flowing over one another at differentspeeds with virtually no mixing between layers,
(2) Fluid particles move in definite andobservable paths or streamlines.
(3) The flow is characteristic of viscous (thick) fluid or isone in which viscosity of the fluid plays a significant part.

- Re < 2000
- 'low' velocity
- Dye does not mix with water
- Fluid particles move in straight lines
- Simple mathematical analysis possible
- Rare in practice in water systems.

TURBULENT FLOW

Turbulent flow is characterized by the irregular

movement of particles of the fluid. There is nodefinite frequency as there is in wave motion. The particles travel in irregular paths with noobservable pattern and no definite layers.

- Re > 4000
- 'high' velocity
- Dye mixes rapidly and completely
- Particle paths completely irregular
- Average motion is in the direction of the flow
- Cannot be seen by the naked eye
- Changes/fluctuations are very difficult to detect. Must use laser.
- Mathematical analysis very difficult - so experimental measures are used
- Most common type of flow.

PRESSURE LOSS

Up to this point on the course we have considered ideal fluids where there have been no losses due to friction or any other factors. In reality, because fluids are viscous, energy is lost by flowing fluids due to friction which must be taken into account. The effect of the friction shows itself as a pressure (or head) loss.

In a pipe with a real fluid flowing, at the wall there is a shearing stress retarding the flow, as shown below.

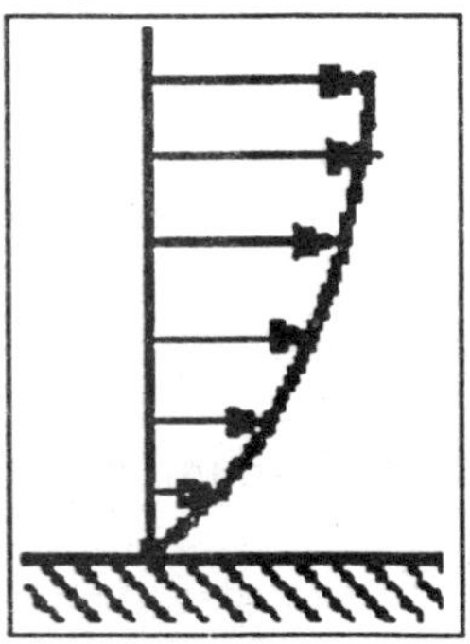

If a manometer is attached as the pressure (head) difference due to the energy lost by the fluid overcoming the shear stress can be easily seen. The pressure at 1 (upstream) is higher than the pressure at 2.

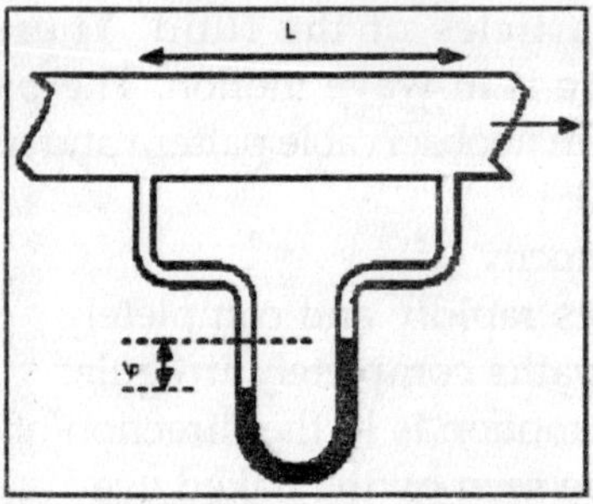

We can do some analysis to express this loss in pressure in terms of the forces acting on the fluid. Consider a cylindrical element of incompressible fluid flowing in the pipe, as shown,

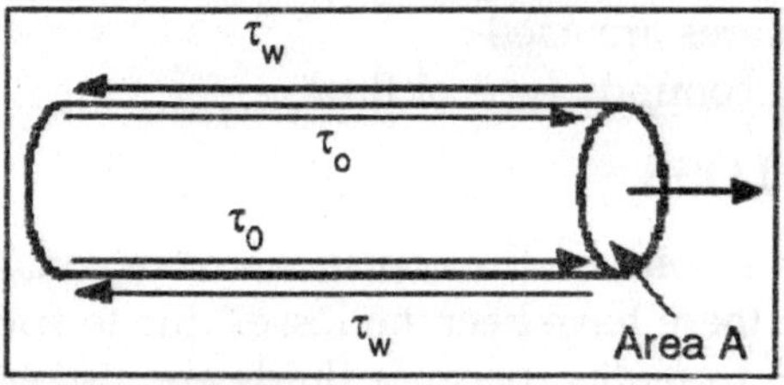

The pressure at the upstream end is *p*, and at the downstream end the pressure has fallen by Δp to (p-Δp). The driving force due to pressure (F = Pressure x Area) can then be written driving force = Pressure force at 1 - pressure force at 2,

$$pA-(p-\Delta p)A=\Delta p A=\Delta p\frac{\pi d^2}{4}$$

The retarding force is that due to the shear stress by the walls,

= shear stress × area over which it acts

= τ_w × area of pipe wall

= τ_w πdL

As the flow is in equilibrium, driving force = retarding force,

$$\Delta p\frac{\pi d^2}{4}=\tau_w \pi dL$$

$$\Delta p=\frac{\tau_{w4L}}{d}$$

Giving an expression for pressure loss in a pipe in terms

of the pipe diameter and the shear stress at the wall on the pipe.

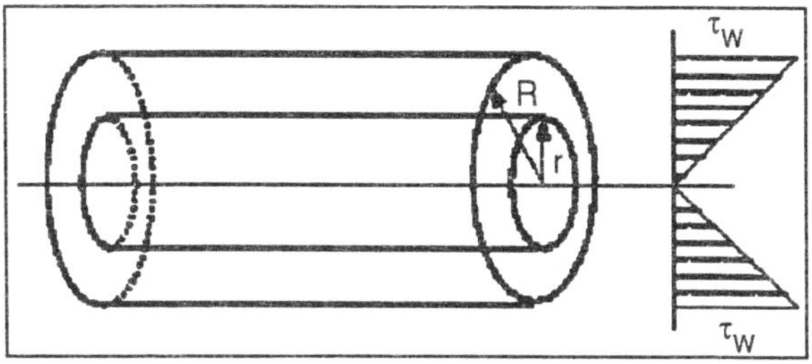

The shear stress will vary with velocity of flow and hence with Re. Many experiments have been done with various fluids measuring the pressure loss at various Reynolds numbers. These results plotted to show a graph of the relationship between pressure loss and Re look similar to the figure below:

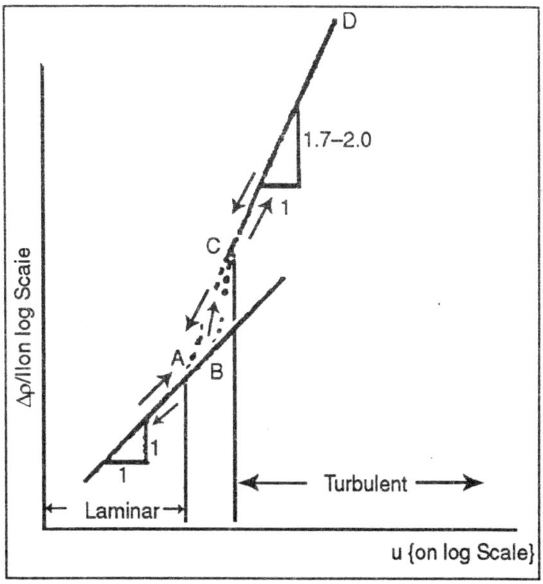

This graph shows that the relationship between pressure loss and Re can be expressed as

laminar $\Delta p \; \alpha \; \mu$

turbulent $\Delta p \; \alpha \; \mu^{17 (\text{or } 2.0)}$

As these are empirical relationships, they help in determining the pressure loss but not in finding the magnitude of the shear stress at the wall τ_w on a particular fluid. If we knew τ_w we could then use it to give a general equation to predict the pressure loss.

PRESSURE LOSS DURING LAMINAR FLOW

In general the shear stress τ_w. is almost impossible to measure. But for laminar flow it is possible to calculate a theoretical value for a given velocity, fluid and pipe dimension.

In laminar flow the paths of individual particles of fluid do not cross, so the flow may be considered as a series of concentric cylinders sliding over each other - rather like the cylinders of a collapsible pocket telescope.

As before, consider a cylinder of fluid, length L, radius r, flowing steadily in the centre of a pipe.

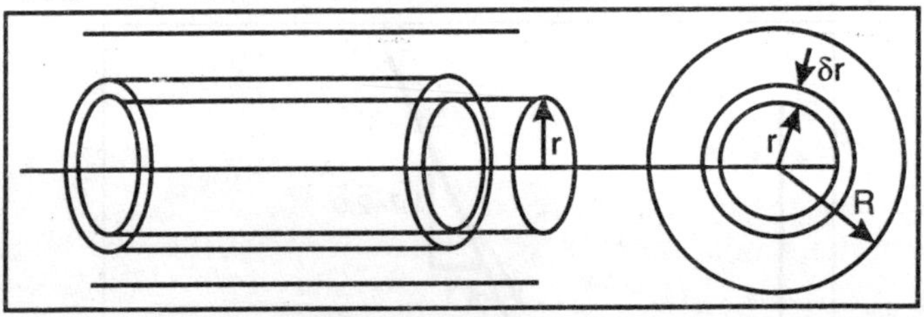

We are in equilibrium, so the shearing forces on the cylinder equal the pressure forces.

$$\tau 2\pi r L = \Delta p A = \pi r^2$$

$$\tau = \frac{\Delta p}{L}\frac{r}{2}$$

By Newtons law of viscosity we have,

$$\tau = \mu \frac{du}{dy},$$

where y is the distance from the wall. As we are measuring from the pipe centre then we change the sign and replace y with r distance from the centre, giving,

$$\tau = \mu \frac{du}{dr}$$

Which can be combined with the equation above to give

$$\frac{\Delta p}{L}\frac{r}{2}=\mu\frac{du}{dr}$$

$$\frac{du}{dr}=-\frac{\Delta p}{L}\frac{r}{2\mu}$$

In an integral form this gives an expression for velocity,

$$u=-\frac{\Delta p}{L}\frac{1}{2\mu}\int r\,dr$$

Integrating gives the value of velocity at a point distance r from the centre,

$$u_r=-\frac{\Delta p}{L}\frac{r^2}{4\mu}+C$$

At $r = 0$, (the centre of the pipe), $u = u_{max}$, at $r = R$ (the pipe wall) $u = 0$, giving

$$C=\frac{\Delta P}{L}\frac{R^2}{4\mu}$$

so, an expression for velocity at a point r from the pipe centre when the flow is laminar is,

$$u_r=\frac{\Delta p}{L}\frac{1}{4\mu}(R^2-r^2)$$

Note how this is a parabolic profile (of the form $y = ax^2 + b$) so the velocity profile in the pipe looks similar to the figure below,

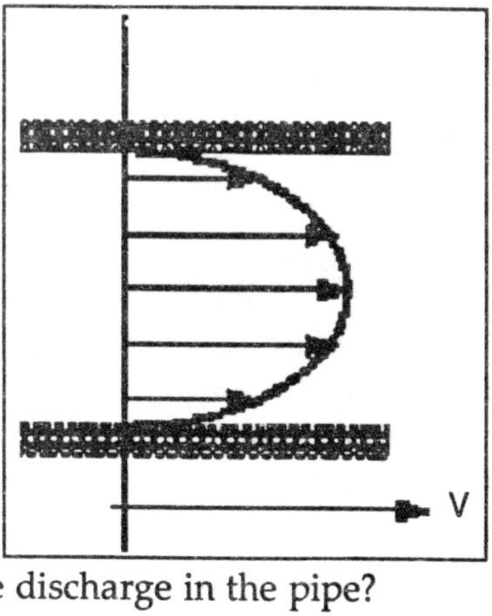

What is the discharge in the pipe?

$$Q = u_m A$$

$$u_m = \int_0^R u_r, dr$$

$$= \frac{\Delta p}{L}\frac{1}{4\mu}\int_0^R u_r\, dr$$

$$= \frac{\Delta p}{L}\frac{R^2}{8\mu} = \frac{\Delta p d^2}{32\mu L}$$

So the discharge can be written,

$$Q = \frac{\Delta p d^2}{32\mu L}\frac{\pi d^2}{4}$$

$$= \frac{\Delta p}{L}\frac{\pi d^2}{128\mu}$$

This is the Hagen-Poiseuille equation for laminar flow in a pipe. It expresses the discharge *Q* in terms of the pressure gradient,

$$\left(\frac{\partial p}{dx} = \frac{\Delta p}{L}\right),$$

diameter of the pipe and the viscosity of the fluid.

We are interested in the pressure loss (head loss) and want to relate this to the velocity of the flow. Writing pressure loss in terms of head loss h_f, i.e. $p = rgh_f$

$$u = \frac{pgh_f d^2}{32\mu L}$$

$$h_f = \frac{32\mu L u}{pgd^2}$$

This shows that pressure loss is directly proportional to the velocity when flow is laminar.

It has been validated many time by experiment.

It justifies two assumptions:

- Fluid does not slip past a solid boundary
- Newtons hypothesis.

ROTATIONAL AND IRROTATIONAL FLOW

From now on, we will only deal with steady, incompressible, inviscid flow. We will neglect all body forces, even gravity.

Previously we have derived an expression for vorticity vector:

$$\vec{\xi} = \vec{D} \times \vec{V}$$

We mentioned that vorticity is a measure of the angular velocity of the fluid elements.

A flow in which there is a lot of vorticity is called a rotational flow. In rotational flows, the fluid elements will rotate as they move from upstream to downstream. This is like a snow ball tumbling down a steep hill side.

A flow in which there is very little vorticity is called an irrotational flow. The fluid particles may move up and down, laterally etc. However, like boxes placed on an escalator, they will maintain their upright position (whatever it is) as they move from point to point.

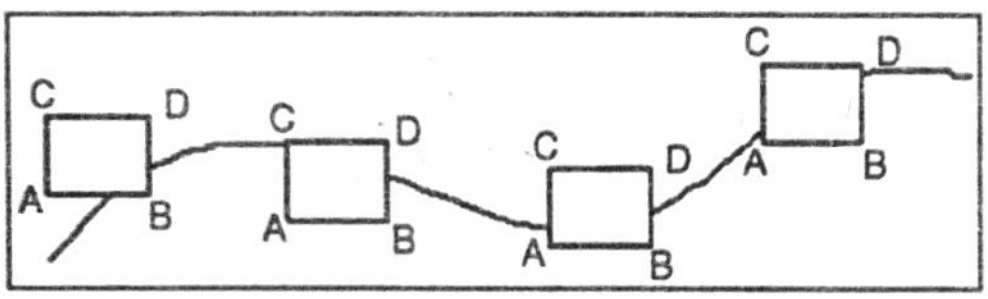

What causes fluid particles to rotate? A number of physical phenomena can.

Phenomena

- Viscous stresses can pull the fluid particles in different direction. Consider a fluid particle right next to a solid surface. The bottom side of the particle may want to stick to the surface due to viscosity. The top side of the particle may be dragged by the particles above in the general direction of the flow. The result is a rotation of the particle.

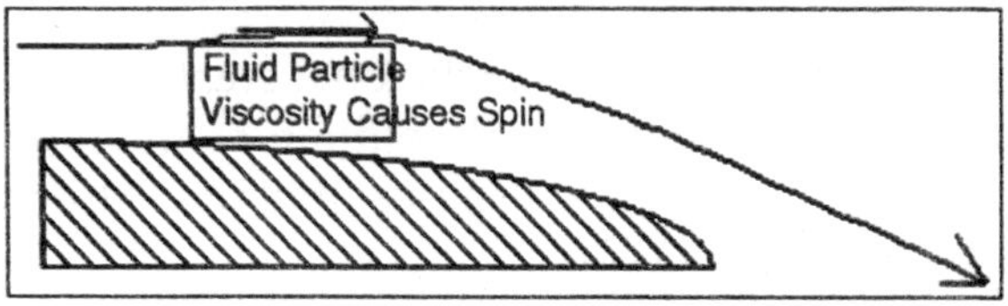

Thus, boundary layers and viscous regions are rotational, and filled with vorticity.

- Velocity differences at the edge of a jet can cause the top part of a fluid particle to move faster than the lower part, or vice versa.

Result: Rotation

- Shock waves can slow down the flow. The strength of shock waves depends on the angle they make with the flow. There is usually a curved "bow shock" ahead of a blunt body. Even though the fluid particles ahead of the shock may be uniform, different fluid particles may impact the shock at different angles, and slowed down differently behind the shock wave.

  Result: Rotation.

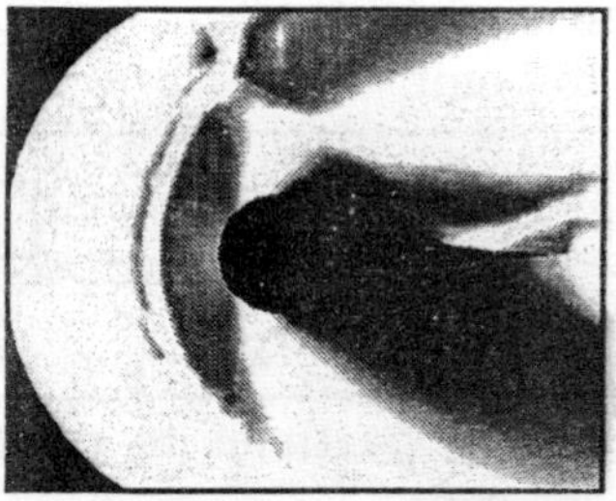

- Non-uniform body forces (example, forces exerted by a spinning propeller on air particles) can cause different fluid particles to be accelerated differently, causing a non-uniform velocity field.

Result: *Rotation*

Although most flows encountered in nature are rotational, if we stay away from viscous regions (boundary.layers), curved shock waves or non-uniform body forces, in all other regions the flow may be assumed to be irrotational.

Assuming the flow to be irrotational greatly simplifies the mathematics, and the eventual solution of the flow on the computer. We will be invoking the irrotational flow assumption soon.

Circulation:

Consider a closed curve C in space. It is made of tiny straight line segments 'ds' each which may be thought of as a short vector. Then circulation over this contour C is defined as.

$$\Gamma = \oint_{c} \vec{V} \cdot d\vec{s}$$

There is a theorem called Stokes Theorem which states that the circulation is related to the vorticity vectors crossing the surface A enclosed by the closed curve C as follows:

$$\Gamma = \oint_{c} \vec{V} \cdot d\vec{s} = \iint_{s} \vec{\xi} \cdot d\vec{A}$$

The area A may be assumed to be made of several infinitesimal elements dA. The vector $d\vec{A}$ has the magnitude of the area dA, and is directed normal to dA.

Stokes theorem says that if there is circulation, then there is vorticity, and vice versa. There will be no circulation over a contour C if the area captured by that contour does not contain any vorticity.

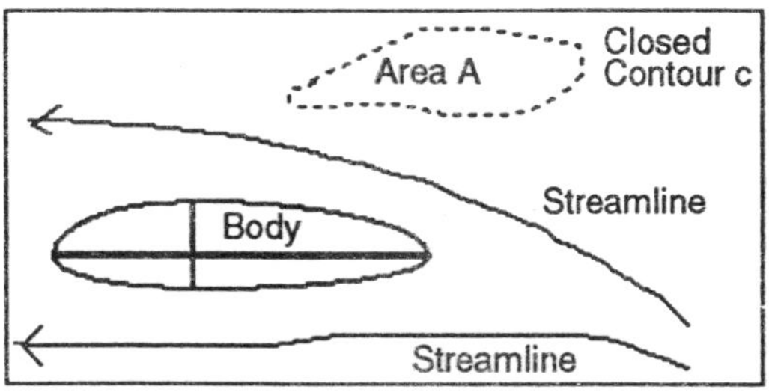

What happens if the contour C encloses the body? Then, since the body has a boundary layer, and the boundary layer contains vorticity, the circulation over any contour that encloses the body will usually be nonzero. The exception is a symmetric body (e.g. symmetric airfoil) which contains clockwise vorticity on one side, say the upper side boundary layer and counterclockwise vorticity on the lower side.

Potential Flow

A irrotational flow (that is, a flow in which vorticity is

zero) is also called a potential flow. This is because we can define a function called the velocity potential F.

If u, v and w are defined as derivatives of Φ as shown, then vorticity in that flow is zero.

This is because,

$$\vec{\xi}=\vec{\nabla}\times\vec{V}=\begin{vmatrix} \vec{i} & \vec{j} & \vec{k} \\ \frac{\partial}{\partial x} & \frac{\partial}{\partial y} & \frac{\partial}{\partial z} \\ \frac{\partial \phi}{\partial x} & \frac{\partial \phi}{\partial y} & \frac{\partial \phi}{\partial z} \end{vmatrix}$$

$$\vec{i}\left(\frac{\partial^2\phi}{\partial y\partial z}-\frac{\partial^2\phi}{\partial zdy}\right)+\vec{j}\left(\frac{\partial^2\phi}{\partial z\partial x}-\frac{\partial^2\phi}{\partial xdz}\right)+\vec{k}\left(\frac{\partial^2\phi}{\partial x\partial y}-\frac{\partial^2\phi}{\partial ydx}\right)=0$$

If we use equation in continuity equation, we get,

$$\vec{\nabla}\cdot\vec{V}=0$$

$$\text{Or,}\left(\vec{i}\frac{\partial}{\partial x}+\vec{j}\frac{\partial}{\partial x}+\vec{k}\frac{\partial}{\partial z}\right)\cdot\left(\vec{i}\frac{\partial \phi}{\partial x}+\vec{j}\frac{\partial \phi}{\partial x}+\vec{k}\frac{\partial \phi}{\partial z}\right)=\frac{\partial^2\phi}{\partial x^2}+\frac{\partial^2\phi}{\partial z^2}=0$$

or,

Ñ²f = 0

The operator ∇^2 is called the Laplacian operator.

It is simply three second derivatives added together as shown below:

$$\nabla^2=\frac{\partial^2}{\partial x^2}+\frac{\partial^2}{\partial y^2}+\frac{\partial^2}{\partial z^2}$$

Equation above is called Laplace's equation for the velocity potential Φ. Why did we introduce a new variable called Φ? It is because we would rather solve a single linear PDE for Φ (given by equation 7) than solve 4 nonlinear PDEs - conservation of mass, u– v–, w-momentum etc. If we can somehow solve equation, we can find u, v and w from equation. Finally, we can find p from the Bernoulli equation.

We will spend much of our remaining time in this course solving Bernoulli equation. This simple equation may be used to model flow over airfoils, wings, bodies etc. It is truely useful.

Of course, it works only for inviscid, incompressible, irrotational flows. It can not be used in regions of large vorticity, e.g. boundary layers.

Stream Function

A less useful new variable is called stream function y. It exists only in 2-D flows, or axi-symmetric flows (e.g. flow around a body of revolution such as a sphere or a bullet).

The stream function is defined as:

$$u = \frac{\partial \psi}{\partial y}$$

$$v = \frac{\partial \psi}{\partial x}$$

We define u and v this way to automatically satisfy the continuity equation for 2-D incompressible flows.

$$\frac{\partial u}{\partial x} + \frac{\partial u}{\partial y} = 0$$

For 2-D flows we only need to worry about the z-component of vorticity given by,

$$\frac{\partial v}{\partial x} - \frac{\partial u}{\partial y}.$$

You may recall that we derived this component by considering the angular velocity of a fluid element in a previous hand-out titled "Pathlines, Streamlines, Vorticity and Bernoulli's Equation". In irrotational flow, vorticity is zero. Setting,

$$\frac{\partial v}{\partial x} - \frac{\partial u}{\partial y}$$

to zero, and using the definitions given in equation we get:

$$\tilde{N}^2 y = 0$$

The above equation is Laplace's equation for stream function. It is linear, and easy enough to solve. After we somehow solve for y we can get u and v from equation. We can finally compute p from Bernoulli's equation.

Equation of a Streamline

Recall that a streamline is defined as:

$$\frac{dx}{u}=\frac{dy}{v}$$

or,

$$udy - vdx = 0$$

Use the definitions of u and v from equation. Then equation becomcs:

$$\frac{\partial\psi}{\partial y}dy+\frac{\partial y}{\partial x}\partial x=0$$

The above equation implies that y is constant along a streamline.

IRROTATIONAL FLOW AROUND A JOUKOWSKI AIRFOIL

Flows of incompressible irrotational fluids are often encountered early in courses on fluid mechanics and PDEs. These flows involve a velocity potential which satisfies Laplace's equation.

There is a rich literature associated with the use of conformal transformations to solve these problems by mapping a simple domain, in which it is easy to solve Laplace's equation, into a more complex domain.

The velocity potential is then interpreted in the new coordinates, thus giving the solution for irrotational flow in the complex geometry.

From the standpoint of CFD, irrotational flow has interesting features. Firstly, Laplace's equation is very much easier to solve than the Navier-Stokes equations. Also, irrotational flow often represents a good approximation to high Reynolds number flows outside boundary layers, provided that the boundary layers have not separated from the boundaries. Another use of irrotational flow solutions is as a starting point for in-depth CFD computations, be they for laminar or turbulent flow.

A key variable used in irrotational flows is the complex potential,

$$w = f + iY$$

where ϕ and Ψ are respectively the velocity potential and streamfunction. The velocity components are given by derivatives of ϕ and Ψ:

$$\mu=-\frac{\partial\phi}{\partial x}=-\frac{\partial\Psi}{\partial y}$$

$$v=\frac{\partial\Psi}{\partial x}=\frac{\partial\phi}{\partial y}$$

The particular complex potential,

$$\omega(\zeta)=V(\zeta+\alpha^2/\zeta)+ik\log\frac{\zeta}{\alpha}$$

represents an irrotational flow consisting of streaming motion past a cylinder of radius α augmented by circulation of strength k around the cylinder.

The transformation defined by,

$$z = z + I^2/z$$

is known as the Joukowski transformation, and maps a circle in the ζ-plane into an airfoil shape in the z-plane. The airfoil is given an angle of attack α by allowing I to be complex, I = Ir + iIi. For later use, define $\zeta = \xi + i\eta$ and note that,

$$\frac{l^2}{\zeta}=\frac{(l_1\,\zeta\, l_2\,\eta)-i(l_1\, h-l_2\,\zeta)}{\zeta^2+\eta^2}$$

where $l_1=l_r^2-l_i^2$ and $l_2=2\,l_r\,l_i$.

To describe irrotational flow around the Joukowski airfoil, we first use *Fastflo* to describe a circle in the ζ-plane. This is a straightforward application for *Fastflo*'s triangular mesh generator. We then use mapmesh to transform from the ζ-plane to the z-plane, and then evaluate contour lines for the streamfunction in the z-plane.

The files for this problems are airfoil.msh and airfoil.prb. The mesh file gives a circle of unit radius (boundary segment 2) located inside a rectangular duct (boundary segment 1). A fictitious segment with tag 0, which has to be crossed twice, forms a bridge between the circle and the duct.

The parameters in this file are circ, which is the circulation k of the flow around the airfoil, c1 and c2, which give x the y coordinates of the re-positioned circle, and ellr and elli, which are the real and imaginary parts of I.

You will find that the airfoil shape is sensitive to small changes to the parameters. In particular, elli is related to the angle of attack and c2 affects the bend. Two other properties susceptible to small changes are the fore-aft asymmetry and the thickness of the airfoil.

The macro run is straightforward. It calls macro conform, assembles problem strf and then displays the streamfunction on the graphics screen. conform and strf are examined in more detail below.

Under the Joukowski transformation, the unit circle in the ζ-plane is first positioned with respect to the origin in the z-plane. This requires two parameters, c1 and c2. The first two lines of conform re-position the mesh and calculate $\zeta^2\eta^2$ as mentioned earlier. The next two lines calculate the parameters I_1 and I^2.

Then,

- V101=X1+(ell1*V401+ell2*V402)/V201
- V102=X2–(ell1*V402–ell2*V401)/V201

complete the calculation of the Joukowski-transformed variable z,adjusted so that the airfoil shape is centred on the origin of the ζ-plane.

```
mapm
black
prim
```

The mapmesh command exchanges the contents of vectors V101 and V102 with the x- and y-coordinates of the mesh. The screen is cleared and the transformed mesh is shown.

```
A   strf
e   -D_jD_jU1=0
b   2 curvature=-0.2071
b   2 U1={0.0}
b   1 U1={V102-V102
       (V100@V100)+circ*log(V100@V100)}
```

A favourable feature of the boundary conditions needs to be pointed out. Because the command mapmesh in conform exchanges V100 with the mesh data, V100 in strf is based on the original mesh data. This means that the boundary condition at tag 1 (the walls of the duct) is evaluated exactly, and there is no need to allow for perturbations to this boundary condition caused by the airfoil.

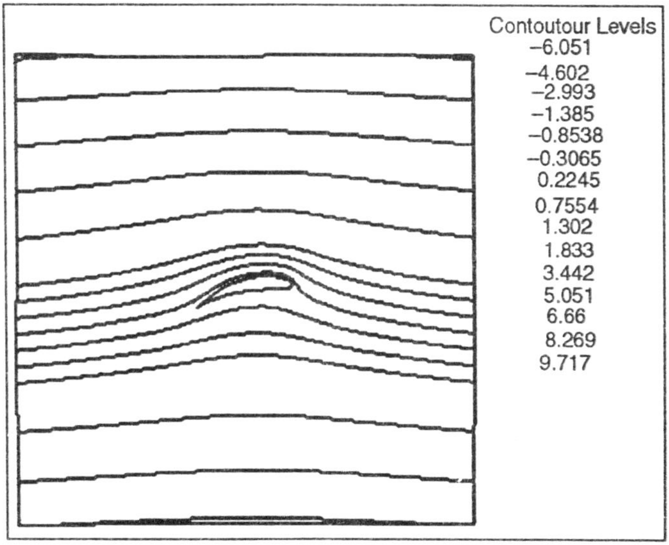

That completes the description of the Joukowski airfoil problem. On loading the files in the usual way, you will find that the problem runs quickly and gives results as shown below. It is instructive to experiment with the parameters, particularly the circulation k.

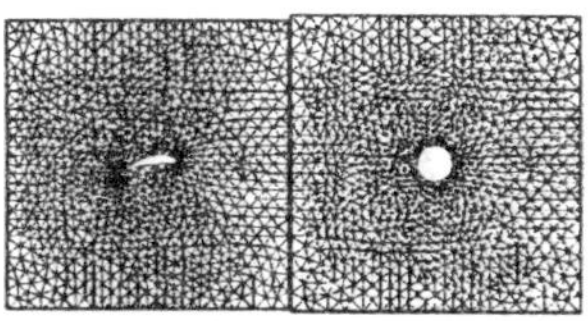

IRROTATIONAL FLOW OF INVISCID FLUID

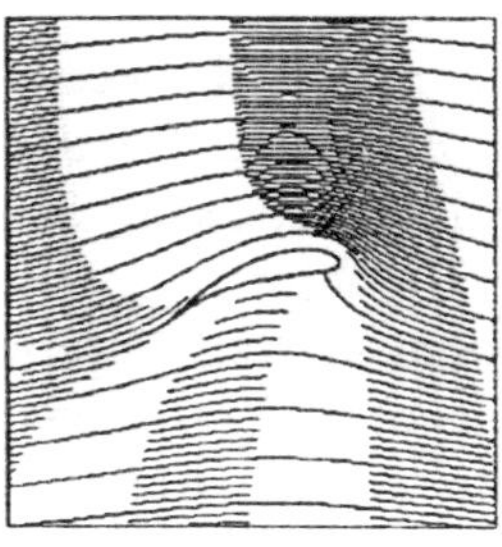

This lecture is devoted to the study of irrotational plane flows of an inviscid fluid. The ultimate goal is to determine the force exerted by a uniform flow on a Joukowski wing (airfoil). It should be noted that, as in a wind tunnel, the force would be the same if the wing moves with the same velocity in a fluid at rest.

In the absence of shear stresses, forces exerted on a body - the wing in this case - by an inviscid fluid are always normal to the surface and due to the pressure. The force on the body is obtained through the determination of the flow and pressure fields on the surface of the object.

It will be shown that this force is always directed at a right angle with respect to flow (lift), that it is proportional to the fluid velocity far from the body and to the circulation around the profile. The component of the force in the direction of the flow (drag) is always zero. This apparent paradox can only be solved by dropping the hypothesis of irrotational (and inviscid) flow in a region close to the body (boundary layer) where viscosity plays a fundamental role and the flow is rotational and viscous.

COMPRESSIBLE AND IN COMPRESSIBLE FLOW

COMPRESSIBLE FLOW

This book deals with an introduction to the flow of compressible substances (gases). The main difference between compressible flow and almost incompressible flow is not the fact that compressibility has to be considered. Rather, the difference is in two phenomena that do not exist in incompressible flow.

The first phenomenon is the very sharp discontinuity (jump) in the flow in properties. The second phenomenon is the choking of the flow. Choking is when downstream variations don't effect the flow.

Though choking occurs in certain pipe flows in astronomy, there also are situations of choking in general (external) flow. Choking is referred to as the situation where downstream conditions, which are beyond a critical value(s), doesn't affect the flow.

The shock wave and choking are not intuitive for most people. However, one has to realise that intuition is really a condition where one uses his past experiences to predict other situations. Here one has to learn to use his intuition as a tool for future use. Thus, not only aeronautic engineers, but other engineers, and even manufacturing engineers will be able use this "intuition" in design and even research.

Flow in which density changes are significant. Pressure changes normally occur throughout a fluid flow, and these pressure changes, in general, induce a change in the fluid density. In a compressible flow, the density changes that result from these pressure changes have a significant influence on the flow.

The changes in the flow that result from the density changes are often termed compressibility effects. All fluids are compressible. However, compressibility effects are more frequently encountered in gas flows than in liquid flows.

An important dimensionless parameter in compressible flows is the Mach number, *M*. This is defined by Eq. shown,

$$M = \frac{V}{\alpha}$$

where *a* is the speed of sound and *V* is the velocity of the flow. For a gas, the speed of sound is given by Eq. shown,

$$\alpha = \sqrt{kRT}$$

where *R* is the gas constant, $k = c_p/c_v$, c_p and c_v being the specific heats at constant pressure and constant volume respectively, and *T* is the temperature. If $M < 0.3$ in a flow, the density changes in the flow will usually be negligible; that is, the flow can be treated as incompressible. Compressible flows are, therefore, as a rough guide, associated with Mach numbers greater than 0.3.

When $M < 1$, the flow is said to be subsonic; when $M = 1$, the flow is said to be sonic; when *M* varies from slightly below 1 to slightly above 1, the flow is said to be transonic; and if *M* > 1, the flow is said to be supersonic. When the Mach number is very high, this usually being taken to mean $M > 5$, the flow is said to be hypersonic.

Compressible flows can have features that do not occur in low-speed flows. For example, shock waves and expansion waves can occur in supersonic flows. Another important phenomenon that can occur due to compressibility is choking, where the mass flow rate through a duct system may be limited as a result of the Mach number being equal to 1 at some point in the flow.

Another effect of compressibility is associated with the acceleration of a gas flow through a duct. In incompressible flow, an increase in velocity is associated with a decrease in the cross-sectional area of the duct, this in fact being true as long as $M < 1$.

However, when $M > 1$, that is, when the flow is supersonic, the opposite is true; that is, an increase in the velocity is associated with an increase in the cross-sectional area.

Therefore, in order to accelerate a gas flow from subsonic to supersonic velocities in a duct, it is necessary first to decrease the area and then, once the Mach number has reached 1, to increase the area, that is, to use a so-called convergent-divergent nozzle. An example is the nozzle fitted to a rocket engine.

COMPRESSIBLE FLOW

Compressible flow appears in many natural and many technological processes. Compressible flow deals with more than air, including steam, natural gas, nitrogen and helium, etc. For instance, the flow of natural gas in a pipe system, a common method of heating in the u.s., should be considered a compressible flow. These processes include the flow of gas in the exhaust system of an internal combustion engine, and also gas turbine, a problem that led to the Fanno flow model. The above flows that were mentioned are called internal flows. Compressible flow also includes flow around bodies such as the wings of an airplane, and is considered an external flow.

These processes include situations not expected to have a compressible flow, such as manufacturing process such as the die casting, injection molding. The die casting process is a process in which liquid metal, mostly aluminum, is injected into a mold to obtain a near final shape. The air is displaced by the liquid metal in a very rapid manner, in a matter of milliseconds, therefore the compressibility has to be taken into account.

Clearly, Aero Engineers are not the only ones who have to deal with some aspect of compressible flow. For manufacturing engineers there are many situations where the compressibility or compressible flow understating is essential for adequate design. For instance, the control engineers who are using pneumatic systems use compressed substances. The cooling of some manufacturing systems and design of refrigeration systems also utilizes compressed air flow knowledge. Some aspects of these systems require consideration of the unique phenomena of compressible flow.

Traditionally, most gas dynamics (compressible flow) classes deal mostly with shock waves and external flow and briefly teach Fanno flows and Rayleigh flows (two kind of choking flows). There are very few courses that deal with isothermal flow. In fact, many books on compressible flow ignore the isothermal flow. In this book, a greater emphasis is on the internal flow. This doesn't in any way meant that the important topics such as shock wave and oblique shock wave should be neglected. This book contains several chapters which deal with external flow as well.

INCOMPRESSIBLE FLOW

Many liquids are hard to compress (i.e., their volumes or densities don't change much when you squeeze them), so that the density ρ is essentially a constant. For such an incompressible fluid the equation of continuity simplifies to,

$$\nabla \cdot \mathrm{v} = 0 \quad \textit{(incompressible}$$

so that the velocity field v is a solenoidal vector. This condition is analogous to the condition $\nabla \cdot \mathrm{B} = 0$ which we encounter in electromagnetism—the magnetic field has zero divergence.

Incompressible flows are much easier to treat theoretically than compressible flows, so this is a very useful simplifying assumption. You might worry about applying it to gases, but it turns out that even there it often works well. In fact, if U is some characteristic flow velocity, and c is the speed of sound in the fluid, then the fluid can be treated as incompressible whenever U<< c. This is often expressed in terms of the Mach number Ma, defined as,

$$Ma = U/c;$$

then if Ma << 1 the fluid can be treated as incompressible. For instance, the speed of sound in air is 330 m/s (740 mi/h); therefore, fluid flow around people, cars, trains, as well as fluid flow in hurricanes and tornadoes, all satisfy the incompressibility condition. Air flow around commercial airliners (Ma = 0.7) violates this condition, as does supersonic flight.

A flow is said to be incompressible if the density of a fluid element does not change during its motion. It is a property of the flow and not of the fluid. The rate of change of density of a material fluid element is given by the material derivative,

$$\frac{D\rho}{Dt} = \frac{\partial \rho}{\partial t} + uj\frac{\partial \rho}{\partial x_j}$$

From the continuity equation we have,

$$\frac{D\rho}{Dt} = -\rho\frac{\partial u_j}{\partial x_j}$$

Hence the flow is incompressible if the divergence of the velocity field is identically zero. Note that the density field need not be uniform in an incompressible flow. All that is required is that the density of a fluid element should not change in time as it moves through space. For example, flow in the ocean can be considered to be incompressible even though the density of water is not uniform due to stratification.

Compressible flow can with good accuracy be approximated as incompressible for steady flow if the Mach number is below 0.3.

Dimensional analysis

Assume L is the characteristic length, U is the

characteristic velocity, the magnitude of velocity gradient is U/L. The physical meaning of the incompressible flow is,

$$\frac{\partial u_j}{\partial x_j} \approx 0$$

or

$$\left|\frac{\partial u_j}{\partial x_j}\right| = \left|\rho^{-1}\frac{d\rho}{dt}\right| << U/L.$$

It is know that,

$$\rho^{-1}\frac{d\rho}{dt} = \rho^{-1}\frac{d\rho\, dp}{dp\, dt} = (pc^2)^{-1}\frac{dp}{dt},$$

where C is the sound speed. Hence, the incompressible condition is,

$$\left|\frac{1}{\rho c^2}\frac{dp}{dt}\right| << U/L.$$

In general, for the liquid, ρ = const and,

$$\frac{\rho}{dt} = 0.$$

So the question is that under what contion, the gas flow can be taken as incompressible flow?

We have,

$$\frac{dp}{dt} = \frac{\partial p}{\partial t} + u_i\frac{\partial p}{\partial x_i} = \frac{\partial p}{\partial t}$$

$$-\rho u_i\left(\frac{\partial u_i}{\partial t} + u_j\frac{\partial u_i}{\partial x_i}\right) + \rho u_i f_i + u_i\frac{\partial \sigma_{i,j}}{\partial x_j}$$

where f_i is the body force (taken as gravitational force in the present document), $\sigma_{i,j}$ is the stress tensor defined as,

$$s_{i,j} = he_{kk}d_{ij} + 2me_{ij},$$

where,

$$e_{ij} = \frac{1}{2}\left(\frac{\partial u_i}{\partial x_j} + \frac{\partial u_i}{\partial x_i}\right).$$

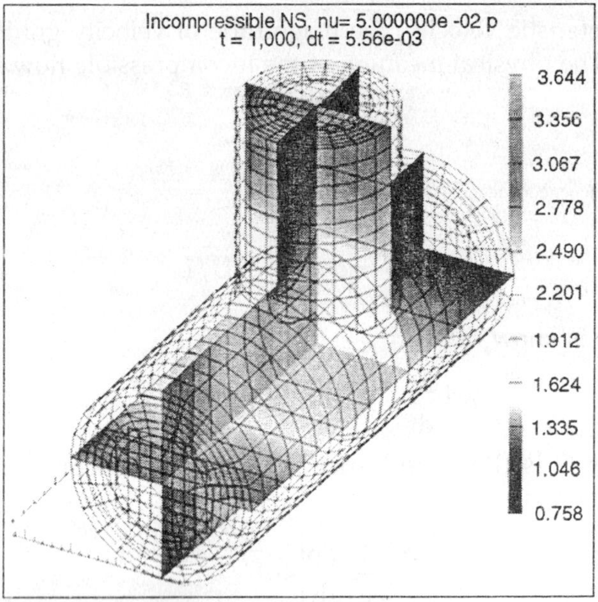

SUBSONIC FLOW AND SUPER SONIC FLOW

SUBSONIC FLOW

Flow of a fluid, as air over an airfoil, at speeds less than acoustic velocity. aerodynamic problems of subsonic flow are treated with the assumption that air acts as an incompressible fluid.

Flow of a fluid at a speed less than that of the speed of sound in the fluid. Also known as subcritical flow.

The screening of sound in a subsonic flow by a cylindrical airbubble layer and a semi-infinite tube

The problem here under discussion lies in the field of sound waves in layered media. The presence of a layer with a velocity of sound less than that of the surroundings will enable sound waves to travel along great distances.

In this domain many investigations have been made e. g. in connection with the propagation of sound in the ocean. Sound velocity in sea-water is a function of depth and reaches a minimum at approximately 1 kilometre under the surface of the sea. It is this minimum that sees to it that waves starting on this level can stay approximately at this level.

In this way sound waves can travel over great distances on a depth which is almost exclusively limited to the direct neighbourhood of this minimum and that is why one is speaking of a sound channel. For the analogous rotational problem Sparenberg has examined the existence of such waves.

SUPERSONIC FLOWS

Many of the techniques used to delay transonic drag rise also are directly applicable in designing the airplane to fly with minimum wave drag in the supersonic regime.

A bow shock wave will exist for free-stream Mach numbers above 1.0. In three dimensions, the bow shock is in reality a cone in shape (a Mach cone) as it extends back from the nose of the airplane. As long as the wing is swept back behind the Mach cone, there is subsonic flow over most of the wing and relatively low drag. A delta wing has the advantage of a large sweep angle but also greater wing area than a simple swept wing to compensate for the loss of lift usually experienced in sweepback. But, at still higher supersonic Mach numbers, the Mach cone may approach the leading edge of even a highly swept delta wing. This condition causes the total drag to increase rapidly and, in fact, a straight wing (no sweep) becomes preferable.

Sweepback has been used primarily in the interest of minimizing transonic and supersonic wave drag. At subsonic Mach numbers, however, the disadvantages dominate. They include high induced drag (due to small wing span or low aspect ratio), high angles of attack for maximum lift, and reduced effectiveness of trailing-edge flaps. The straight-wing airplane does not have these disadvantages. For an airplane that is designed to be multimission, for example, cruise at both subsonic and supersonic velocities, it would be advantageous to combine a straight wing and swept wing design. This is the logic for the variable sweep or swing-wing.

Although not necessarily equal to the optimum

configurations in their respective speed regimes, it is evident that an airplane with a swing-wing capability can, in a multimissioned role, over the total speed regime, be better than the other airplanes individually. One major drawback of the swing-wing airplane is the added weight and complexity of the sweep mechanisms. But technological advances are solving these problems also.

In addition to low-aspect-ratio wings at supersonic speeds, supersonic wave drag may also be minimized by using thin wings and area ruling. Also, long, slender, cambered fuselages minimize drag and improve the spanwise lift distribution.

The SST

President John F. Kennedy committed the United States to "develop at the earliest practical date the prototype of a commercially successful supersonic transport superior to that being built in any other country in the world...." What lay ahead was years of development, competition, controversy, and ultimately rejection of the supersonic transport (SST) by the United States.

The National Aeronautics and Space Administration (NASA) did considerable work, starting in 1959, on basic configurations for the SST. There evolved four basic types of layout that were studied further by private industry. The aircraft manufacturer Lockheed chose to go with a fixed-wing delta design; whereas another aircraft company, Boeing, initially chose a swing-wing design.

One problem associated with the SST is the tendency of the nose to pitch down as it flies from subsonic to supersonic flight. The swing-wing can maintain the airplane balance and counteract the pitch-down motion. Lockheed needed to install canards (small wings placed toward the airplane nose to counteract pitch down). Eventually, the Lockheed design used a double-delta configuration and the canards were no longer needed. This design proved to have many exciting aerodynamic advantages.

The forward delta begins to generate lift supersonically

(negating pitch down). At low speeds the vortices trailing from the leading edge of the double delta increase lift. This means that many flaps and slats could be reduced or done away with entirely and a simpler wing design provided. In landing, the double delta experiences a ground-cushion effect that allows for lower landing speeds. This is important since three-quarters of airplane accidents occur in takeoff and landing. The British-French Concorde and the Russian Tupolev Tu-144 prototypes use a variation of the double delta wing called the ogee wing. It, too, uses the vortex-lift concept for improvement in low-speed subsonic flight.

Ultimately, Boeing with a swing-wing design was selected as the winner of the U.S. SST competition. The size of the Boeing SST design grew to meet airline payload requirements. Major design changes were incorporated into the Boeing 2707-100 design. The supersonic cruise lift-drag ratio increased from 6.75 to 8.2, and the engines were moved farther back to alleviate the exhaust impinging on the rear tail surfaces.

Despite the advantages previously quoted for a swing-wing concept, technological advances in construction did not appear in time. Because of the swing-wing mechanisms and beefed-up structure due to engine placement, incurable problems in reduction of payload resulted. Boeing had no recourse but to adopt a fixed-wing concept—the B2707-300. Political, economic, and environmental factors led the United States to cancel the project in 1972.

While the British-French Concorde and Russian Tu-144 fly, research is still continuing into advanced supersonic transports in the United States. Whereas, the Concorde and Tu-144 cruise at Mach = 2.2 to 2.4, and the Boeing design cruised at Mach = 2.7, configurations with a cruise speed of Mach = 3.2 have been being analysed.

Sonic Boom

One of the more objectionable of the problems facing any supersonic transport is commonly referred to as the "sonic boom." To explain sonic boom, one must return to a description of the shock-wave formation about an airplane flying supersonically. A typical airplane generates two main shock waves, one at the nose (bow shock) and one off the tail (tail shock). Shock waves coming off the canopy, wing leading edges, engine nacelles, etc. tend to merge with the main shocks some distance from the airplane.

The resulting pressure pulse changes appear to be N-shaped. To an observer on the ground, this pulse is felt as an abrupt compression above atmospheric pressure followed by a rapid decompression below atmospheric pressure and a final recompression to atmospheric pressure. The total change takes place in one-tenth of a second or less and is felt and heard as a double jolt or boom.

The sonic boom, or the overpressures that cause them, is controlled by factors such as airplane angle of attack, altitude, cross-sectional area, Mach number, atmospheric turbulence, atmospheric conditions, and terrain. The overpressures will increase with increasing airplane angle of attack and cross-sectional area, will decrease with increasing altitude, and first increase and then decrease with increasing Mach number.

Turbulence in the atmosphere may smooth the "N" wave profile and thus lessen the impact of the boom or, on the other hand, may in fact amplify the overpressures. Reflections of the overpressures by terrain and buildings may cause multiple booms or post-boom aftershocks. In a normal atmospheric profile, the speed of sound increases with decreasing altitude.

The directions in which the overpressures travel are refracted in this normal case and they will at some point curve away from the Earth. The strongest sonic boom is felt directly beneath the airplane and decreases to nothing on either side of the flight path. It is interesting to note that a turning supersonic airplane may concentrate the set of shock waves locally where they intersect the ground and produce a superboom. Perhaps the greatest concern expressed about the sonic boom is its effect on the public. The effects run from structural damage (cracked building plaster and broken windows) down to heightened tensions and annoyance of the citizenry. For this reason, the world's airlines have been forbidden to operate supersonically over the continental

United States. This necessitates, for SST operation, that supersonic flight be limited to overwater operations. Research for ways in which to reduce the sonic boom continues.

CR!TICAL AND SUPER CRITICAL FLOW

CRITICAL FLOW

The model is based on broadcasting or dissemination information of 'critical' value (which by its very nature will not be disclosed by those involved with bad governance practices) to targeted audience using ICT and convergent media.. Targeted audience may include, media, opposition parties, judicial bench, independent investigators or the wider public domain itself.

The use of this model requires a foresight of:

- Understanding the "critcal and use value" of a particular information set.
- How or from where this information could be obtained.
- How could the information be used strategically.
- Who are the best target group for such information- the users for whom the availability of this information will make a huge difference.

The strength of this model is that the concept of 'distance' and 'time' becomes redundant when information is hosted on a digital network. Once available on the digital network, the information could be used advantageously- by instantly transferring the critical information to its user group located anywhere or by making it freely available in the wider public domain.

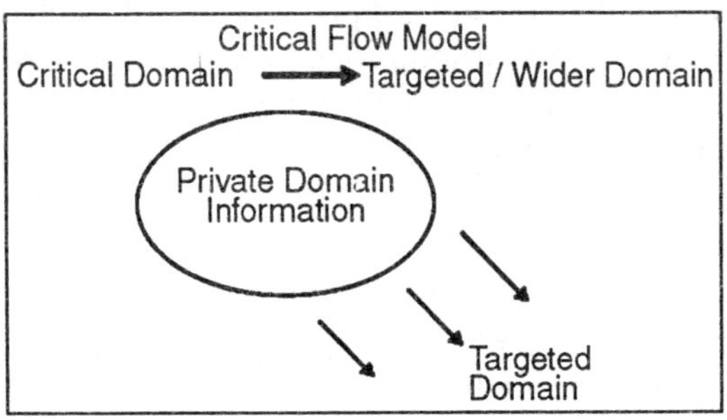

Applications

This model could be applied in the following possible ways:

- Making available Corruption related data about a particular Ministry / Division/ Officials online to its electoral constituency or to the concerned regulatory body.
- Making available Research studies, Enquiry reports, Impact studies commissioned by the Government or Independent commissions to the affected parties.
- Making Human Rights Violations cases violations freely available to Judiciary, NGOs and concerned citizens.
- Making available information that is usually suppressed, for instance, Environmental Information on radioactivity spills, effluents discharge, information on green ratings of the company to concerned community.

SOUND SPEED AND CRITICAL FLOW

Critical flow (sometimes called "choked flow") is important for many industrial applications involving multiphase flow. These applications include the design of throttling valves for refrigeration systems, bypass systems for steam power plants, emergency relief in chemical and nuclear plants and other diverse applications.

Most of the research and development in the recent past has been in regard to plant safety analyses. Our discussion of critical flow is based on experimental and model results with this latter subject as the main application, although the models are completely general and applicable to other design issues.

One of the observations made in examining multiphase pressure drop was that one could determine if the multiphase flow could be considered compressible or incompressible by examining the acceleration terms in the pressure gradient expression.

For instance, in the HEM model for the pressure gradient it was noted that the equation denominator could be considered as$(1 - M^2)$ where M is a local two-phase flow Mach number; i.e., ratio of mass velocity to a two-phase "acoustic" mass velocity. Analogous to single phase flow, as the Mach number approaches unity the flow must be considered compressible and the overall pressure gradient increases; asymptotically approaching an infinite gradient (shock wave) as the Mach number equals one.

Thus, as in single phase flow, there is a relationship in this multiphase flow model between the maximum mass flux through a duct and the local sound speed. This maximum mass flux is identified as the critical flow (or choked flow") of the system, and is the major topic of this section.

These notes will begin with a discussion of single phase sound speed and critical flow ("choking"). This logically leads us to consider the homogeneous equilibrium model for sound speed and for the multiphase critical flow.

Finally, more sophisticated models considering nonequilibrium effects and the effects of phase velocity differences are discussed. We end the paper with a discussion of these models assumptions and recommendations for usage.

In this development, as in the discussion of pressure drop, we assume the flow is approximately one-dimensional. This simplifying assumption allows us to develop analytical expressions and is approximately valid in realistic situations if the change in flow area is not abrupt.

Many practical applications can involve at least two-dimensional flow. At worst the one-dimensional assumption affords one a way to estimate behaviour in more complex multi-dimensional systems to bound the expected behaviour.

Understanding Critical Flow

The concept of critical flow is not an easy one to impart, but some understanding of it is necessary to understand the full range of flow conditions that may exist in a reach. We start by understanding that water has appreciable internal cohesion.

As water accelerates, that is it moves away from water flowing more slowly behind it, it does not initially separate, but rather stretches, as the forces of cohesion, including gravity and viscosity, are being overcome by the increasing movement of the water. This can be seen as water passes over a weir or approaches a constriction in the channel (e.g. a flume).

The water falling over the crest of the weir is accelerating and indeed may break up, but as it does so its store of kinetic energy increases until it is released in a chaotic explosion of activity. This increase in kinetic energy can be seen by a noticeable dip in the water surface just upstream of the crest of the weir. Prior to this dip the velocity of water is said to be sub-critical; at the dip it is critical and below the dip, it is super-critical.

Stream flow cannot go on accelerating in the natural river environment. At some stage their needs to be a resolution of super-critical flow, as deceleration occurs and the water assumes the same velocity as the water in front of it (sub-critical velocity). It does so by means of a hydraulic jump, where the super-critical flow, on reaching the slow moving water at the base of the weir, riffle, set of rapids, or individual rock or log, is swept upwards and falls back on itself, like a crashing wave, but falling in the upstream direction.

Here the water decelerates and returns to its sub-critical state, to join the slower, placid, steadily moving downstream flow, unless of course there is yet another fall to be negotiated. Over the course of the hydraulic jump undissipated kinetic energy is explosively converted into turbulence and potential energy which lifts the water level after the jump. Whether a section of channel will yield a change in (critical) flow, from sub-critical to super-critical, can be assessed by calculating the Froude number.

Essentially the Froude number compares the velocity of flow with what is called the critical velocity for the reach - the point at which water appears to stretch out. A value of one is the point of critical velocity, which itself is simply the square root of the product of acceleration due to gravity multiplied by the depth of the reach.

Depth and gravity are the key factors in determining critical flow. Where the Froude number is less than 1 the water flow has no opportunity to accelerate beyond the critical velocity. It is hindered by the bulk of water immediately downstream. The flow is considered sub-critical. Turbulence can and does exist but local velocities are not sufficient to overcome the internal viscous character of the water. At a Froude number greater than 1 the water flow has been able to accelerate (over a hydraulic drop) and has now exceeded the critical.

Its energy state is unstable and its flow is considered super-critical. The sudden enforcement of sub-critical conditions on this flow as it re-encounters the inertia of the water downstream, results in a release of internal kinetic energy, some of which overcomes the cohesive forces holding the water molecules together foaming the water and tossing it about as can be seen in rapids

SUBCRITICAL AND SUPERCRITICAL FLOW

A flow can be classified as subcritical or supercritical by comparing the ratio of inertial and gravitational forces at a stream location. The inertial forces are characterized by the velocity term (V^2), and the gravitational forces are represented by the term gD.

The ratio of the inertial forces to the gravitational forces is called the Froude number:

$$Fr = \frac{V}{\sqrt{gD}}$$

where,

Fr	=	the Froude Number (dimensionless)
V	=	the average velocity (ft/s, m/s)
g	=	the gravitational constant (32.2 ft/s^2, 9.81 m/s^2)
D	=	the hydraulic depth (ft, m)

(2.9)

When Fr > 1, the flow is supercritical and inertial forces dominate. As a result, the channel velocity is high and the depth is low, with the flow described as rapid or shooting. Supercritical flow is generally associated with steeper slopes. Flow in a street gutter is often supercritical, due to the usual shallow depths of a few tenths of a ft (several cm) combined with a velocity of 1–2 ft/s (0.3–0.6 m/s). For Fr < 1, the flow is said to be *subcritical,* with gravitational forces dominant.

Consequently, the flow has a relatively low velocity and high depth, and it may be described as calm or tranquil. This type of flow is generally associated with small channel slopes and is the most common type of flow in natural channels. However, a stream may have an average velocity of 10 ft/sec (3 m/s) and the flow would be subcritical if the hydraulic depth were 3.3 ft (1 m) or more. For Fr = 1, both the depth and the flow are said to be *critical.*

Critical flow is a transitional condition, where neither inertial nor gravitational forces dominate. Only a small change in velocity or depth causes the flow to change to subcritical or supercritical. Critical depth is further addressed in Section 2.6. The real-world condition is normally one of subcritical flow, and most river systems have Froude numbers less than 0.5, even during major floods. Similarly, man-made channels normally have subcritical flow.

Supercritical flow occurs most often in man-made channels; however, steep mountain streams can be supercritical, especially during floods. Figure 2.16a shows a steep, natural stream in Switzerland that appears to be in supercritical flow.

Examples of supercritical flow include flow in street gutters during a rainfall event, flow down a spillway, and flow in steep concrete channels. Log rides in amusement park flumes are also good examples of supercritical flow.

Fig. (a) Supercritical Flow in a Natural Channel and (b) Flood Bore, Unsteady and Rapidly Varied, in the Same Channel.

Steady, uniform, gradually varied, or rapidly varied flow conditions for subcritical or supercritical flow regimes may be adequately addressed with the procedures outlined in this book. However, some flow situations require special methods outside the scope of this book. Figure 2.16b shows the same scene just a few minutes after the picture in Figure 2.16a was taken. This high-velocity flood wave (flood bore) is similar to what is expected from a dam-break flood event. For this flood wave, the discharge increased from about 175 to 21,200 ft^3/s (5 to 600 m^3/s) in a matter of seconds. Figure above.16b thus represents an unsteady, rapidly varied flood event.

Special analysis procedures, beyond the scope of this book, are required to estimate the speed and height of the leading edge of the flood wave. The movement of a dam-break type flood can be simulated with unsteady flow modeling, however.

Example:Computing Hydraulic Variables

Water flows at a depth (y) of 4 ft in a trapezoidal channel with a bottom width (B_o) of 10 ft and side slopes of 1V:3H (z = 3). If the discharge is 400 ft^3/s, determine the velocity, hydraulic depth, and Froude number.

Solution: The cross-sectional area of flow is the area of the trapezoidal-shaped channel:

$A = (B_o + zy)y = (10 + 3 \times 4)4 = 88\ ft^2$

The top width of flow is

$T = Bo + 2zy = 10 + (2)(3)(4) = 34\ ft$

The hydraulic depth is

$D = A/T = 88/34 = 2.59\ ft$

The velocity is

$V = Q/A = 400/88 = 4.55\ ft/s$

The Froude Number is

$Fr = V/(gD)^{0.5} = 4.55/[(32.2)(2.59)]^{0.5} = 0.50 < 1$

Therefore, the flow is subcritical.

Example:Computing Hydraulic Variables

If the Froude number is 1.2 for the same hydraulic depth as Example, compute the velocity and discharge in the channel.

Solution: The new values are

$Fr = 1.2 = V/(g \times 2.59)^{0.5}$

$V = 10.96$ ft/s

$Q = (10.96)(88) = 964.4$ ft^3/s

A much steeper slope would be required to pass this higher discharge, compared to the slope that resulted in the flow of 400 ft^3/s.

SIGNIFICANCE OF SUPER CRITICAL FLOW

Super critical flow conditions have both ecological and hydrological implications of great significance to stream management.

For the stream ecosystem the hydraulic jump is a very significant component. Firstly it entrains air into the water column, oxygenating the stream ecosystem. Secondly, it provides a 'launch pad' for fish seeking to leap obstructions in their efforts to move upstream. For jumps that have a chute form, water may also swirl upstream either side of the fall of water, again assisting aquatic organisms to move upstream.

Thirdly, the hydraulic jump is the only time that the pure movement of water makes sound (the babbling of brooks and the roaring of rapids), through the bursting of bubbles, as the entrained air rises to the surface. And fourthly, the super-critical flow conditions are primary habitat for certain aquatic organisms. For modified or human-made channels supercritical flow has advantages and disadvantages.

It is not uncommon for engineers to create channels with sufficient depth, smoothness and slope to create super-critical flow; in order to move water faster through a small narrow channel, and thus free-up land for development or reduce the amount of land that needs to be purchased. Such channels however are inherently dangerous and must be fenced-off. They must also be made of concrete or stone to prevent erosion from the powerful flows.

Super critical flow, with its high velocity and turbulent motion, often with standing waves and occurring on a slippery surface, presents a considerable hazard to people and other terrestrial fauna. Water, only a foot or two deep, but moving at a super critical velocity can sweep a person or animal downstream and hold them under water, as canoeists well know.

The creation of super critical flow is not always intended. Channel modifications that result in a shorter than natural channel, such as meander cut-offs or channel straightening, can create a channel of very steep slope sufficient to cause super critical flow. The effect is enhanced when the new channel is made particularly smooth by desnagging or vegetation clearance. Needless to say, should the channel be formed in non-cohesive material, serious erosion will result.

An interesting way of observing super-critical flow is by throwing a stone into flowing water and observing the ripples. If the ripples move upstream from the point of emanation then the flow is sub-critical, but if the ripples are swept downstream, then the flow is super-critical. Even better compare the rippling effect of objects projecting above the water at points upstream and downstream of critical flow. No ripples will travel upstream of the latter.

ONE, TWO AND THREE DIMENSIONAL FLOW

ONE-DIMENSIONAL FLOW EXAMPLES

Find

Pressure Head, Elevation Head and Total Head, Seepage Velocity

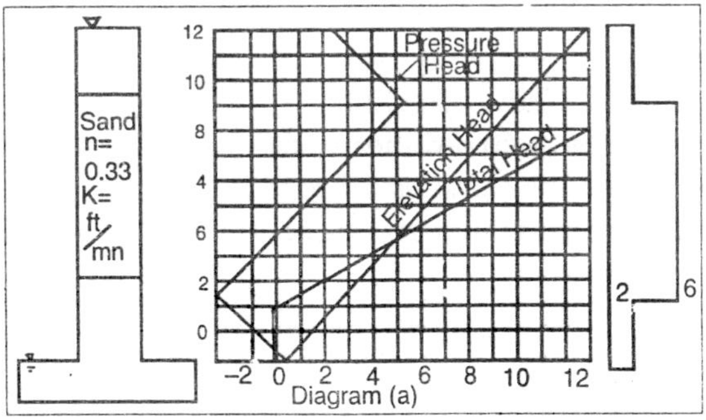

Diagram (a)

THREE DIMENSIONAL FLOW

Although in general all fluids flow three-dimensionally, with pressures and velocities and other flow properties varying in all directions, in many cases the greatest changes only occur in two directions or even only in one. In these cases changes in the other direction can be effectively ignored making analysis much more simple.

Flow is *one dimensional* if the flow parameters (such as velocity, pressure, depth etc.) at a given instant in time only vary in the direction of flow and not across the cross-section. The flow may be unsteady, in this case the parameter vary in time but still not across the cross-section.

An example of one-dimensional flow is the flow in a pipe. Note that since flow must be zero at the pipe wall - yet non-zero in the centre - there is a difference of parameters across the cross-section.

Should this be treated as two-dimensional flow? Possibly - but it is only necessary if very high accuracy is required. A correction factor is then usually applied.

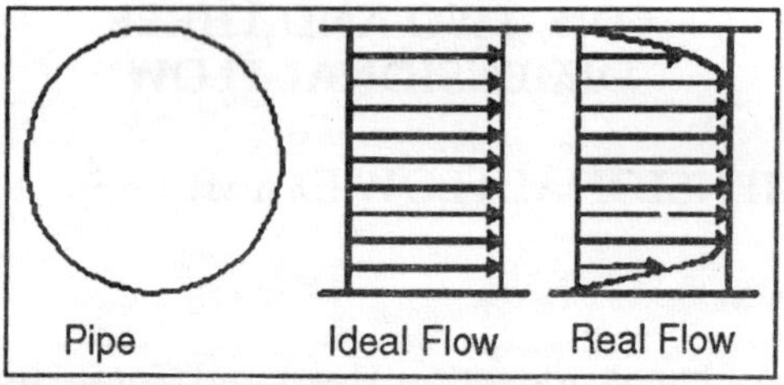

Fig. One-dimensional Flow in a Pipe.

Flow is *two-dimensional* if it can be assumed that the flow parameters vary in the direction of flow and in one direction at right angles to this direction.

Streamlines in two-dimensional flow are curved lines on a plane and are the same on all parallel planes. An example is flow over a weir foe which typical streamlines can be seen in the figure below.

Over the majority of the length of the weir the flow is the same - only at the two ends does it change slightly. Here correction factors may be applied.

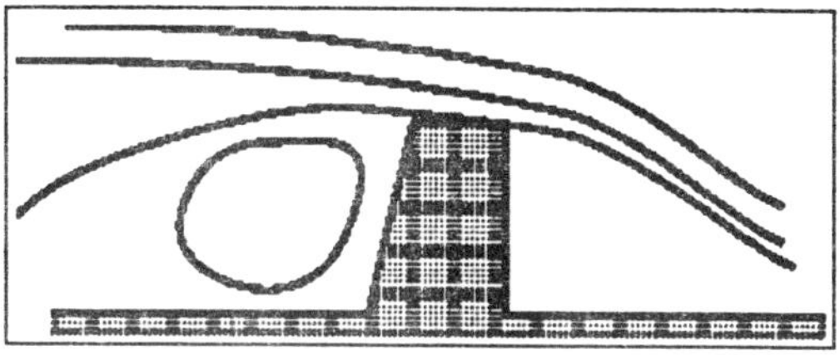

Fig. Two-dimensional Flow Over a Weir.

In this course we will only be considering steady, incompressible one and two-dimensional flow.

THREE DIMENSIONAL FLOW FIELDS

As teraflop and petaflop computers become increasingly accessible, large-scale simulations that can produce three-dimensional, time-dependent scalar and vector data begin to play an instrumental role in problem solving and scientific discovery. For the past two decades, researchers have developed various visualization techniques to enable effective analysis of scientific data. Among the many research directions, visualization of vector fields has remained one of the most active research topics in the field.

Generally speaking, two popular types of methods are used to visualize 2D/3D vector fields. The first is to synthesize textures that follow the flow directions everywhere in the field. One example of such techniques is the Line Integral Convolution (LIC) algorithm [1]. Although LIC was originally designed for visualizing 2D vector fields, various 3D extensions of LIC have also been proposed [2][3]. Figure 1 shows examples of 2D and 3D flow textures generated by extensions of LIC.

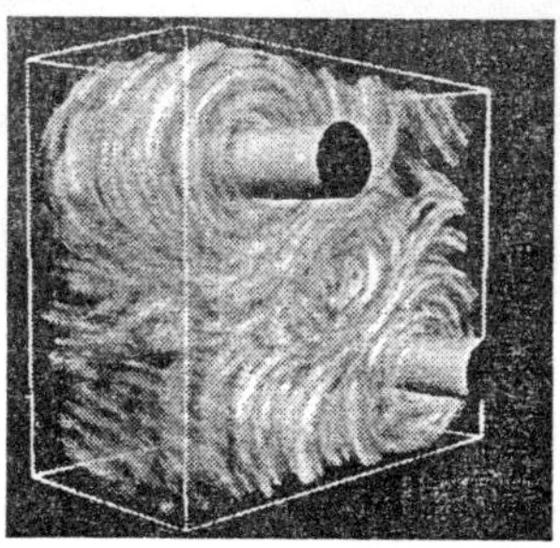

Fig. Visualizing 2D and 3D Vector Fields Using Flow Textures

The other method for visualizing flow data is to insert mass-less particles to the field and observe their trajectories computed with numerical integration. This is sometimes referred to as particle tracing.

While straightforward, this is still a more popular approach because the output resembles what the scientists are used to seeing in traditional experimental flow visualizations. From the particle trajectories, geometric objects such as streamlines, stream ribbons, or stream surfaces can be created and visualized. If it is a time-varying flow field, the particle traces can be used to construct pathlines, streaklines, or time lines. Figure 2 shows an examples of numerical flow visualization.

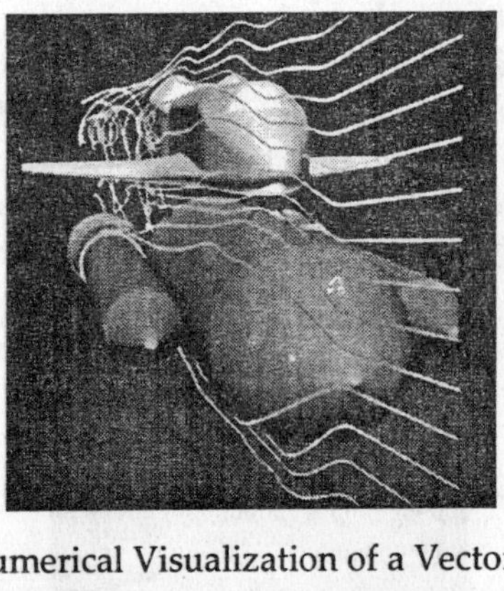

Fig. Numerical Visualization of a Vector Field.

Since streamlines are still widely used by application scientists, visualization researchers have put a lot of effort into making streamline visualization easier and more effective. In this article, I focus on an important issue of visualizing vector fields using streamlines - how to interact with 3D vector field data and place streamline seeds so that the final visualization is insightful and free of clutter.

There have been several seed placement algorithms proposed in the past. A general goal of this research is to generate enough numbers of streamlines to show important flow features without creating visual clutter. To date, most of the available methods are two dimensional, with very little work done for 3D data.

Among the existing 2D algorithms, one common strategy is widely used to reduce visual clutter That strategy is to generate streamlines that are evenly distributed in space so that streamlines are not too close to each other. While those algorithms work reasonably well for 2D data, extending the 2D seed placement algorithms to 3D is nontrivial because as the 3D streamlines are projected to the screen, they can arbitrarily intersect or overlap with one another. The final image can easily become very cluttered if too many streamlines are displayed.

To tackle the 3D streamline placement problem, let's first consider how artists draw. Starting from a blank canvas, strokes are placed one at a time in selected local regions, which are often the projection of the 3D objects that the artists have in mind to draw; for example, the surface of a river, or the body of the cloud.

Depending on the desired effect, the distance between the strokes is carefully controlled. The key observation here is that even though the objects intended to be drawn are three dimensional, all the strokes are placed and controlled on the canvas, which is a 2D space. Figure shows a famous sketch of water flow by Leonardo da Vinci.

Inspired by the process described above, we have recently developed an image-based streamline placement algorithm to tackle the problem of visualizing 3D streamlines. The major goal of the method is to let the user directly interact with the 3D vector data on the screen. Another goal of the algorithm is to tightly couple the exploration of flow data with visualization of other related variables produced from the same simulation.

As we know, multiple variables generated from a simulation are often correlated. To define the region of interest, the user can employ standard visualization techniques such as isosurfaces or slicing planes. These are computed from some flow related quantity such as velocity magnitude, flow density, energy, pressure, etc. The user can interactively visualize the data by changing the iso-value or the orientation of the slicing plane. When a region of interest is found, the user can query the flow directions in the region with our algorithm described blow.

To place streamlines, we first read back the depth values stored in the depth buffer based on what is being visualized on the screen. Then, the user can pick a point on the screen inside the region of interest. From the depth value and the user-picked screen position, we can infer the corresponding 3D position in object space, and drop a seed there. From the seed, we compute a streamline based on the 3D flow field data, and immediately project the streamline back to the screen. From the first computed streamline, we can pick more seeds on the screen with a preset screen distance away from the existing streamline. As we integrate those streamlines, we keep monitoring the distance of the streamlines to the existing ones in screen space. If they are too close to each other, we immediately stop the integrations so that the final image will not be too cluttered.

With this algorithm, we can generate visualizations of streamlines that are free of visual clutter. Figure below shows an isosurface of an implicit stream function and the streamlines generated from the surface. Some additional run time control of the appearance of the streamlines can also be added. For example, we can perform level of detail rendering of streamlines. Assuming we use a constant screen space distance threshold, as we zoom out of the domain, fewer streamlines will be generated because the visualized object now becomes smaller. Figure again shows an example of level of detail rendering of streamlines.

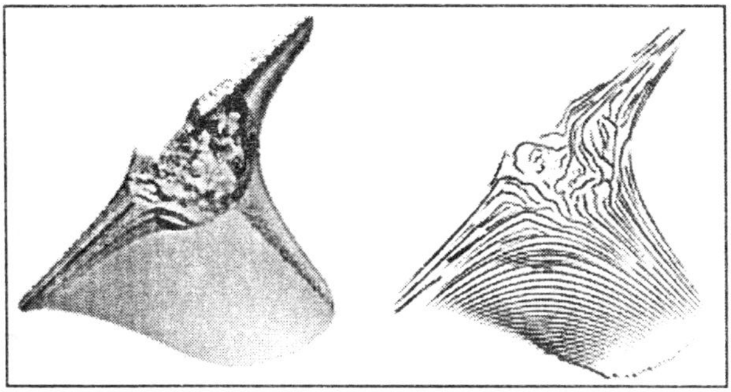

Fig. An isosurface of a Stream Function and the Corresponding *Streamlines.*

We can display streamlines in different depth layers. Since streamline seeds are dropped on the screen and attached to the 3D positions defined by the depth map, we can modify the seed positions easily by increasing or decreasing the depth values from the original depth map. We can also 'cut a region open' by modifying the depth values in the depth map with a mask, and drop seeds there.

STREAMLINE FLOW

DEFINITION

When an object moves through a fluid, there will be friction due to the resistance of the material medium. This resistance increases with the velocity of the body as in the case of a rocket and aeroplane. Hence to minimize this friction due to the resistance of the medium, the bodies of aeroplanes.

In real life an object which is moving over a horizontal surface does not continue with constant velocity when the accelerating force is removed. Instead the object slows down and eventually stops. The deceleration is caused by a frictional force between the object and the surface on which it travels. A frictional force always acts to oppose the motion of an object over a surface and is an example of a contact force.

When we try to slide one object over another, we find that there is a force which opposes the motion. This force is called friction.

Whenever there is relative motion of two surfaces in

contact, there is frictional resistance. When a body moves over another, it has to overcome friction. This needs energy. Hence, some energy is always wasted in overcoming friction. When two parts of a machinery rub against each other heat is produced, and the efficiency of the machine decreases. There will be a greater wear and tear in the different parts of a machine due to friction.

If the surfaces that are in contact are rough, there is more friction. Hence to reduce friction, the surfaces in contact are kept smooth and polished. It is worth noting that if the two surfaces in contact are of the same metal, the friction increases on polishing the two surfaces. This is why the bearings and the axle are made of different metals.

HOW IS FRICTION HELPFUL

From what has been discussed above, it may appear that if there were no friction, our life would have been much more easier and we would not have to spend energy in overcoming the frictional resistance. This is not true. As a matter of fact, it is only because of friction that we are able to move around. Try to run on a very smooth surface and see what happens. As we step forward, friction opposes the forward force, so that the foot does not slip forward.

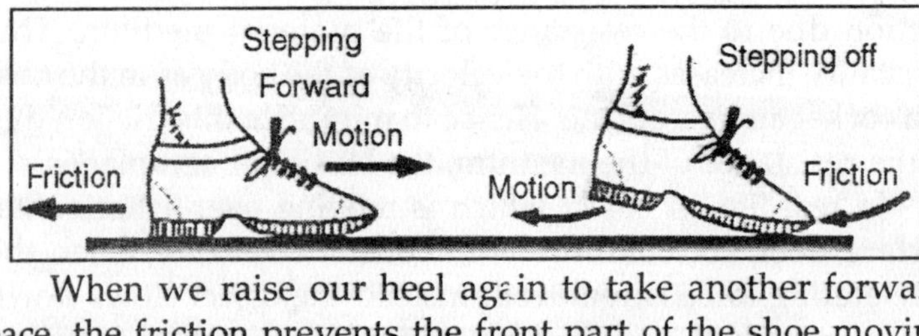

When we raise our heel again to take another forward pace, the friction prevents the front part of the shoe moving back. On account of the friction between the ground and the rubber type of a bicycle, the latter can move forward. In the absence of friction between the wheels and ground, the wheels are unable to roll. This is why the wheels of bicycle and other vehicles are treaded to increase friction.

To stop the bicycle, we use the opposing frictional force between the brake blocks and the wheel rim. When the driver applies the brakes of an automobile, the friction of the brake lining on the brake drum slows down the wheels and enable him to bring the car to a stop. Friction holds a screw in wood. It also enables us to use nails to fix boards together. If friction were eliminated, dishes would slide off the table unless the table were perfectly level. Thus without friction, our life would have been, in reality, very inconvenient.

Types of Friction

You must have noticed that when you have to slide a heavy box, it is easy to slide it once the box has started sliding. You may also have noticed that it is much more easy to roll a heavy drum than to slide it. This is because there are three types of friction.

They are:

- The Static Friction.
- The Kinetic or Dynamic Friction.
- The Rolling Friction.

The friction that exists between the two surfaces in contact when there is no relative motion between them is called static friction.

The force which makes the surface just start to slide over another, is called limiting friction.

The friction that exists between a surface sliding on another surface is called kinetic or dynamic friction.

When a body rolls on the surface of another, the friction that exists between the surfaces is known as rolling friction.

In the example described above, when the box just starts moving, the friction is the static friction. Once it has started moving, the friction that exists is the sliding friction. In the case when the drum is rolling on the surface, the friction is the rolling friction.

Experimentally, it has been proved that:

Static friction > Kinetic or Dynamic friction > Rolling friction

Laws of Friction

- Friction always acts in such a direction that it opposes

The motion. Force of friction is directly proportional to normal reaction.

- Until motion takes place, the static friction adjusts itself so as to be equal to the force tending to produce motion.
- In any given case, there is a limiting value to the frictional force.
- Limiting friction is independent of the area of contact.
- Limiting friction is directly proportional to the normal reaction.

Ways to Reduce Friction

Lubricants

Use of lubricants makes the surface smooth. In machines, friction is reduced by using lubricants such as oil, grease or graphite powder. The lubricant fills the minute unevenness of the two surfaces and separates them by forming a very thin layer in between. This layer offers very little resistance and as a result, the friction gets reduced.

Ball Bearings or Roller Bearing

They are made in various designs to reduce friction. They reduce friction because rolling friction is smaller than sliding friction. These bearings are also lubricated to further improve their function.

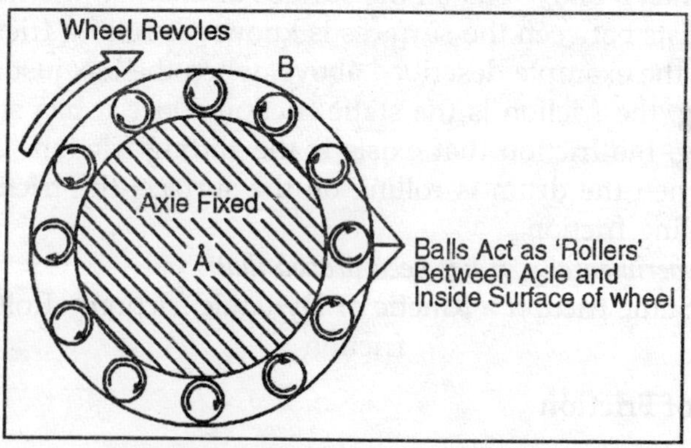

Fig. Ball-bearing and Rolling Friction

Anti-friction metals

When steel slides over an alloy of lead, the friction is less than when steel slides on steel. Bearings are sometimes packed with such alloy.

Air cushion

By separating the surfaces by an air cushion e.g., a hovercraft, can travel over rough land, swamp or sea. Here, as the moving object does not come in contact with the other solid surface, the frictional forces are reduced to a very great extent.

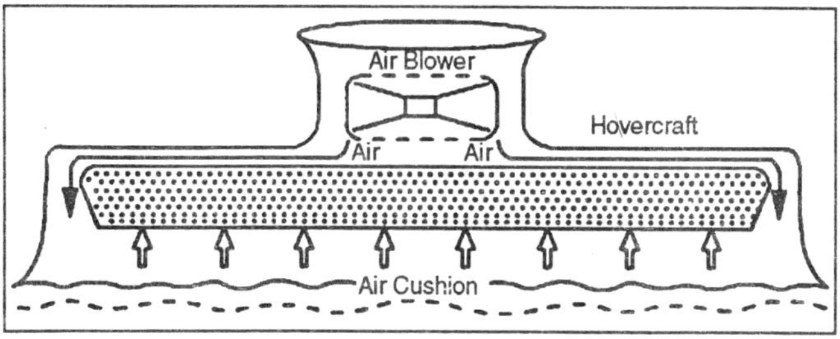

Fig. Hovercraft Principle

Streamline flow

When an object moves through a fluid, there will be friction due to the resistance of the material medium. This resistance increases with the velocity of the body as in the case of a rocket and aeroplane. Hence to minimize this friction due to the resistance of the medium, the bodies of aeroplanes, missiles etc. are streamlined. In nature also, we find the same adaptation of shape to suit the medium, specially in aquatic animals. Compared to air, resistance to movement in water is much more. To overcome this resistance, fish etc. have a shape that offers least resistance. Shapes of the birds also manifest this adaptation.

By Polishing

Polishing the surfaces reduces the ruggedness between them. In an automobile when the piston moves in the cylinder the friction is reduced by polishing the surfaces, which are in contact. The jewel bearing and polished agate edges reduce the friction to a minimum. This is because by polishing the unevenness is reduced to minimum. In a chemical balance the beam rests on an agate knife-edge, where the friction is reduced. Roads are made smooth by concreting and cementing the surface. The speed of a body moving through the fluid also depends upon the size and shape of the body [Figure (a) shown below] shows the streamlines of the fluid through which a body in the shape of a sphere is moving.

Such smooth flow is known as 'Laminar flow' and occurs at relatively low speeds. As the speed increases, the resistance and friction increases. When such a stage is reached, 'turbulence' sets in and the force of friction increases to a great extent. The turbulent flow is accompanied with eddy currents which decreases the speed of the body. As a result, a lot of energy is wasted in overcoming this resistance [Figure (b) shown below]. The shapes of the bodies are designed in such a way that they offer minimum resistance while they move in air or water [Figure (c) shown below]. This reduces the tendency to form eddy currents and the body moves through the fluid with minimum resistance. Such shapes are called streamline shapes and the resultant is called streamline flow.

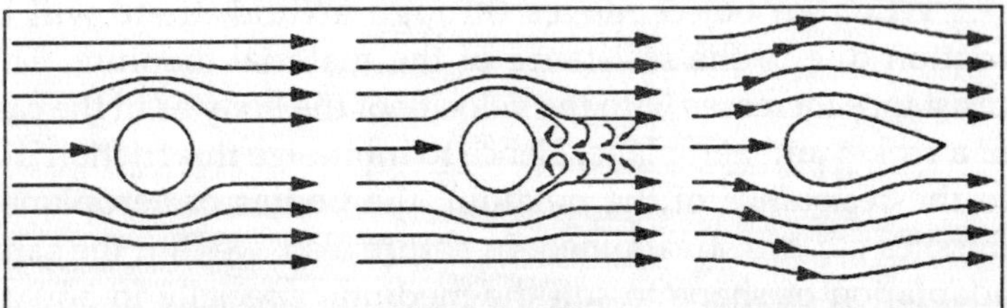

(a) Laminar Flow at Slow Speed — (B) Turbulent Flow at Fast Speed — (C) Streamline Flow at Fast Speed

CONTINUITY EQUATION FOR 3D AND 1D FLOW

CONTINUITY EQUATION

When a fluid is in motion, it must move in such a way that mass is conserved. To see how mass conservation places restrictions on the velocity field, consider the steady flow of fluid through a duct (that is, the inlet and outlet flows do not vary with time). The inflow and outflow are one-dimensional, so that the velocity V and density \rho are constant over the area A.

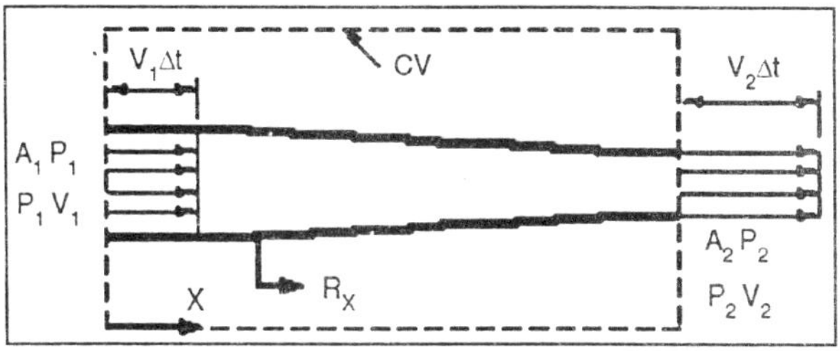

Fig. One-dimensional Duct Showing Control Volume.

Now we apply the principle of mass conservation. Since there is no flow through the side walls of the duct, what mass comes in over A_1 goes out of A_2, (the flow is steady so that there is no mass accumulation). Over a short time interval\Delta t,

$$\text{volume flow out over } A_2 = A_2V_2\Delta t$$

Therefore:

mass in over A = $\rho A_1V_1\Delta t$

mass in over A = $\rho A_2V_2\Delta t$

So: $\rho A_1V_1 = \rho A_2V_2$

This is a statement of the principle of mass conservation for a steady, one-dimensional flow, with one inlet and one outlet. This equation is called the continuity equation for steady one-dimensional flow.

For a steady flow through a control volume with many inlets and outlets, the net mass flow must be zero, where inflows are negative and outflows are positive.

STREAMLINES AND STREAMTUBES

A streamline is a line that is tangential to the instantaneous velocity direction (velocity is a vector, and it has a magnitude and a direction). To visualize this in a flow, we could imagine the motion of a small marked element of fluid. For example, we could mark a drop of water with fluorescent dye and illuminate it using a laser so that it fluoresces.

If we took a short exposure photograph as the drop moves

according to the local velocity field (where the exposure needs to be short compared to the time it takes for the velocity to change appreciably), we would see a short streak, with a length V\Delta t, and with a direction tangential to the instantaneous velocity direction. If we mark many drops of water in this way, the streamlines in the flow will become visible. Since the velocity at any point in the flow has a single value (the flow cannot go in more than one direction at the same time), streamlines cannot cross. except at points where the velocity magnitude is zero, such as at a stagnation point.

There are other ways to make the flow visible. For example, we can trace out the path followed by our fluorescent drop using a long-exposure photograph. This line is called a pathline, and it is similar to what you see when you take a long-exposure photograph of car lights on a freeway at night.

It is possible for pathlines to cross, as you can imagine from the freeway analogy: as a car changes lanes, the pathline traced out by its lights might cross another pathline traced out by an adjoining vehicle at a different time. Another way to visualize flow patterns is by streaklines. A streakline is the line traced out by all the particles that passed through a particular point at some earlier time. For instance, if we issued fluorescent dye continuously from a fixed point, the dye makes up a streakline as it passes downstream. To continue the freeway analogy, it is the line made up of the lights on all the vehicles that passed through the same toll booth. If they all follow the same path (a steady flow), a single line results, but if they follow different paths (unsteady flow), it is possible for the line to cross over itself. In unsteady flow, streamlines, pathlines and streaklines are all different, but in steady flow, streamlines, pathlines and streaklines are identical.

FLUID FLOW AND CONTINUITY EQUATION

Fluids, by definition can flow, but are essentially incompressible. This provides some very useful information about how fluids behave when they flow through a pipe, or a hose. Consider a hose whose diameter decreases along its length, as shown in the Figure below. The "continuity equation" is a direct consequence of the rather trivial fact that what goes into the hose must come out. The volume of water flowing through the hose per unit time (i.e. the flow rate at the left must be equal to the flow rate at the right or in fact anywhere along the hose. Moreover, the flow rate at and point in the hose is equal to the area of the hose at that point times the speed with which the fluid is moving:

flow rate = (area) × (velocity)

You can easily verify that (area) x (velocity) has units m^3/t which is correct for volume per unit time.

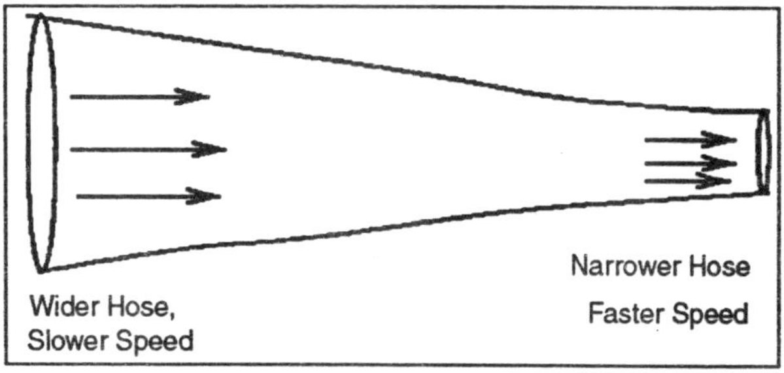

Fig. Fluid Flow in a Hose of Variable Size

These considerations lead us directly to the continuity equation, which states that:

area × velocity = constant

everywhere along the hose. This has the important consequence that as the area of the hose decreases, the velocity of the fluid must increase, in order to keep the flow rate constant.

Anyone who has pinched one end of a garden hose has experienced this effect: the smaller you pinch the end of the hose, the faster the water comes out.

1D AND 3D MODELS OF CEREBROVASCULAR FLOW

The Circle of Willis (CoW) is a ring-like structure of blood vessels at the base of the brain that distributes arterial blood to the cerebral mass. 1D and 3D CFD models of the CoW have been created to simulate clinical scenarios, such as occlusions in afferent arteries and absent circulus vessels. Both models capture cerebral haemodynamic auto-regulation using a Proportional-Integral controller to modify efferent artery resistances to maintain optimal efferent flowrates for a given circle geometry and afferent blood pressure. The models can be used to identify at-risk cerebral arterial geometries and conditions prior to surgery or other clinical procedures. The 1D model is particularly relevant in this instance, with its fast solution time suitable for real-time clinical decisions and scenario testing, as long as it captures the necessary details as a 3D model would.

Results show excellent correlation between models for the transient efferent flux profile with differences less than 5%. The assumption of strictly Poiseuille flow in the 1D model allows more flow through the geometrically extreme communicating arteries than the 3D model. This discrepancy was overcome by increasing the resistance to flow in the ACoA in the 1D model to better match the resistance seen in the 3D model, significantly improving correlation of the results.

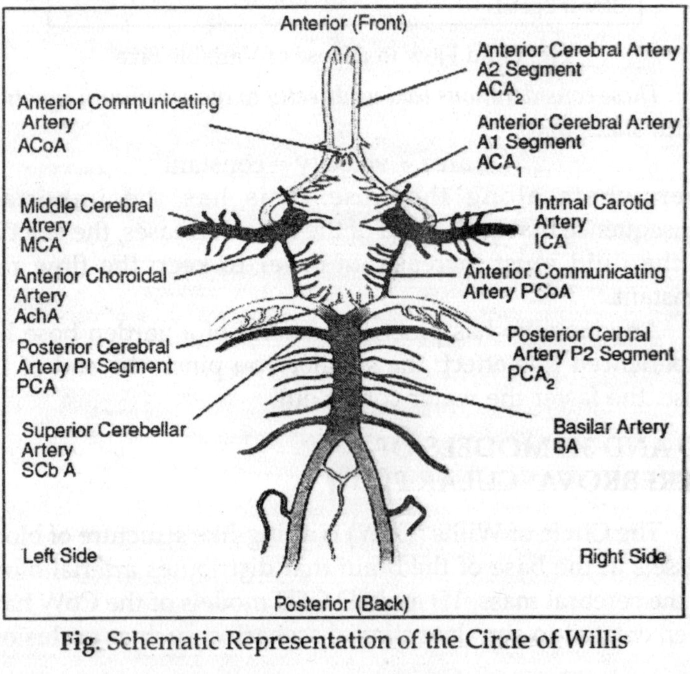

Fig. Schematic Representation of the Circle of Willis

Irrespective of pressure variations in the afferent vessels, the circulus vessels comprising the CoW allow for a constant supply of blood to be delivered to the cerebral mass

MULTILAYER SAINT-VENANT SYSTEM

For some applications like the transport of tracers of different densities or the effect of wind on a lake, we need to know the vertical profile of the velocity and by definition, we cannot get relevant information from Saint-Venant equations. These questions are also fundamental in ocean models and atmospheric sciences.

In these cases, and in order to avoid the 3D Navier-Stokes system when large scale problems are considered, we introduce a new multilayer Saint-Venant system. Thanks to a precise analysis of the shallow water assumption we propose an approximation of the *hydrostatic Navier-Stokes equations* which consists in a set of coupled Saint-Venant systems. It extends the range of validity and gives a precise description of the *vertical profile of the horizontal velocity* while preserving the computational efficiency of the classical Saint-Venant system. The fluid is divided in a given number of layers, each layer satisfies the Saint-Venant equations and is linked to the others by pressure terms (water height coupling) and viscosity (velocity coupling). This approach is a transition step towards 3D simulations with the Navier-Stokes system with two main advantages: we have not to deal with a moving mesh and the vertical velocity is only an output variable, it is not a variable of the system since the kinematic boundary condition is applied at each interface.

The Saint-Venant kinetic solver has been extended to this multilayer model and we have performed a numerical implementation which preserves the stability properties of the kinetic scheme. The pressure source terms of the multilayer system appears in a non conservative form, then this term is divided in two parts, a conservative part introduced in the left hand side of the system and a non consersative part treated as a source term and which has to be approximated carefully to obtain a stable scheme. For the bottom topography term we apply the*hydrostatic reconstruction* to the total water depth and the reconstructed local water depths are deduced.

The 1D and 2D multilayer solvers have been developed and validated. The Figures show the good agreement of the results obtained with the 2D multilayer solver and with the 3D hydrostatic Navier-Stokes solver (presented in the next section) for a transcritical flow over a bump.

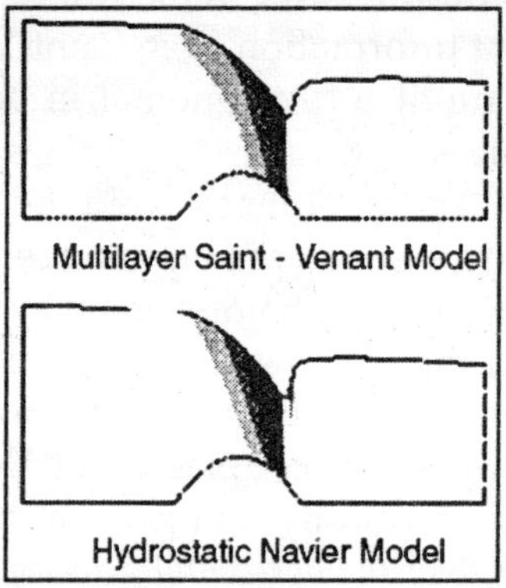

Fig. Transcritical Flow Over a 2D Bump.
Longitudinal cross-section of vertical velocities

3D FREE SURFACE NAVIER-STOKES

Besides this multilayer Saint-Venant approach, we are also studying the 3D incompressible Navier-Stokes equations with free surface. These equations are solved by the Telemac-3D software developed by EDF/LNHE, and our aim is to apply our results to this software in order to improve it.

We have considered the model with *hydrostatic approximation,* which can be viewed either as an independent model or as an intermediate step in the solution of the non-hydrostatic problem. After time discretization of the equations, the fractional step method can be applied in order to split the non-linear advection terms. The system coupling the 3D horizontal velocity and the 2D free surface function is studied.

It is composed by the resulting momentum equation and the depth-integrated continuity equation. We have established a variational formulation, leading to a *mixed and symmetric problem*. We have analysed the semi-discrete problem and its approximation, and proved that they are generally well posed, provided the time step is not too big and the viscosity coefficient not negligible. In the other case, a particular "inf-sup condition" must be satisfied.

This condition is automatically satisfied at the space-continuous level, but not at the discrete level. We have therefore proposed a couple of stable finite elements to approximate the 3D horizontal velocity and the 2D free surface. Finally, we have implemented the resolution of this system into the Telemac-3D system, and the method has been validated through many numerical examples. For instance, we have simulated the three-dimensional flow in an estuary whose geometrical characteristics are very close to those of the Sado estuary in Portugal.

Our aim was to verify that the scheme reproduces the vertical recirculation phenomenon in the region of steep bathymetry making the transition between the estuary and the coastal waters.

We have also worked on improving the treatment of the moving surface in 3D domains. In the Telemac-3D system, the classical *sigma-transformation* technique is applied, which is widely used in the atmospheric and oceanographic communities. The link between this technique and the *ALE approach* was still not clear. Therefore, we have shown that the sigma-transformation is equivalent to the ALE approach for a particular type of ALE mappings. This transformation technique presents many advantages; essentially, the accurate assimilation of the bed and surface boundaries and the possibility of easily incorporating boundary layers.

But the classical variants of the sigma-transformation – which are still the ones mostly used – present important drawbacks. Indeed, they limit the possibilities of vertical discretization and therefore prevent from correctly adapting the mesh to the particular needs of the simulation. Especially, they prevent from using horizontal levels, which is essential to the accurate representation of horizontal density-stratifications. Taking advantage of the reinterpretation of the sigma-transformation as a particular ALE formulation, we propose a very general transformation, allowing for a great adaptability of the vertical discretization and therefore overcoming some drawbacks of the classical sigma-transformation.

In an illustration the improvement obtained using the generalized sigma-transformation for the simulation of a standing basin with a horizontal density-stratification. Figure 5 shows the results obtained using a classical sigma mesh: they reveal a perturbation in the stratification and the creation of spurious velocities.

Finally, we have presented a *monotonic and conservative* numerical scheme for the resolution of the linear advection problem set on a moving domain. The main idea is to use the *Multidimensional Upwind Residual Distributive* (MURD) approach, well-known for its ability to satisfy the properties just mentioned, as well as the high accuracy and stability of the solution. However, the schemes based on this approach have been developed for problems defined on fixed domains, and their extension to *moving domains* is not yet well-established.

It has been proven by several authors that the domain movement can affect numerical schemes in a negative way, leading to a loss of their accuracy, stability or conservation properties. In certain cases, the satisfaction of a particular condition – the so-called *Geometric Conservation Law* (GCL) – ensures the preservation of these properties. We have derived a generic form of MURD schemes for the non-conservative ALE formulation of the continuous linear advection problem with divergence-free advection velocity set on a moving domain.

We have then shown that the properties ensured by the residual distribution schemes set on fixed domains are not altered by the domain movement, except from the mass preservation of the advected quantity. Thus, we have formulated the additional condition an *ALE-MURD scheme* must satisfy to be mass-conservative when the domain moves.

This conservation constraint is strongly linked to the GCL.

Finally, we have applied this conservative advective ALE-MURD scheme to the particular framework of three-dimensional free surface flows.

GRANULAR FLOWS, DEBRIS AVALANCHES

The hypothesis of shallow flow is generally well satisfied by debris avalanches. In collaboration with the Laboratoire de Modélisation et Tomographie Géophysique (IPGP, Paris 7) and the Department of Structural and Geotechnical Engineering, Politecnico di Torino, Italy, we have considered the simulation of debris avalanches using the Saint-Venant model in which the behaviour of granular media is taken into account by a Coulomb type friction term.

We have recently studied the case where the earth pressure coefficient is not constant but depends whether the downslope flow is expanding or contracting. To simulate this effect, we assume that the earth pressure coefficient depends on the sign of the divergence. We show the influence of this varying coefficient on some numerical results.

CIRCULATION

DEFINITION OF CIRCULATION

In fluid dynamics, circulation is the line integral around a closed curve of the fluid velocity. Circulation is normally denoted Γ. *If V is the fluid velocity and ds is a unit vector along the closed curve* C:

$$\Gamma = \oint_C V \cdot ds$$

The dimensions of circulation are length squared over time.

Circulation was first used independently by Frederick Lanchester, Wilhelm Kutta, and Nikolai Zhukovsky. n fluid dynamics, circulation is the path integral around a closed curve of the fluid velocity. Circulation is normally denoted Γ. If is the fluid velocity and the closed curve is denoted C: For a body in an inviscid flow field, lift is equal to the product of the circulation about the body, the air density, and the velocity. Circulation is often used in computational fluid dynamics as an intermediate variable to calculate forces on an airfoil or other body.

CIRCULATION, VORTICITY, AND VORTICES

Fluids will invariably swirl around, and it is often useful to characterize the amount of swirling. This is done by introducing the *circulation* Γ, defined as

$$\Gamma = \oint_C \mathbf{v} \cdot d\mathbf{l},$$

with the line integral taken is around a closed circuit C in a counter-clockwise sense. By using Stokes' theorem we may write this as a surface integral,

$$\oint_C \mathbf{v} \cdot d\mathbf{l} = \int_S (\nabla \times \mathbf{v}) \cdot d\mathbf{S}$$

$$= \int_S \omega \cdot d\mathbf{S},$$

where the *vorticity* ω of the fluid is defined as,

$$w = \nabla \times v.$$

If the vorticity of the fluid is zero, we say that the fluid is irrotational:

$$\nabla \times v = 0 \text{ (irrotational)}.$$

We've already mentioned that there are some useful analogies between magnetostatics and incompressible fluid flow. I've collected these together in Table below. To transform the magnetostatics expressions into those for fluid flow, we make the substitutions,

$$B/\mu_o \rightarrow v,\ I \rightarrow \Gamma, \text{ and } \mu_o \rightarrow \rho$$

(where here ρ is the mass density of the fluid). If a fluid is both incompressible *and* irrotational, then $\nabla \times v = 0$ and $\nabla \cdot v = 0$, which are the same as the equations of electrostatics and magnetostatics in free space (i.e., in the absence of charges and currents).

Table. Analogies Between Incompressible Fluid Flow and Magnetostatics

Magnetostatics	Incompressible fluid flow
$\nabla \cdot B = 0$	$\nabla \cdot v = 0$ $\oint B \cdot dI = \mu_o I$

(Ampére's Law)	$\oint V \cdot dI = \Gamma$
(circulation) $E_B = B^2 / 2\mu_o$	
(energy density of the magnetic field)	$E = \frac{1}{2}\rho v^2$ (kinetic energy density of a fluid)

As an example of fluid flow with vorticity, we can consider a single *vortex*, which would be produced when draining a bathtub, and which is illustrated in Fig. 2.5.

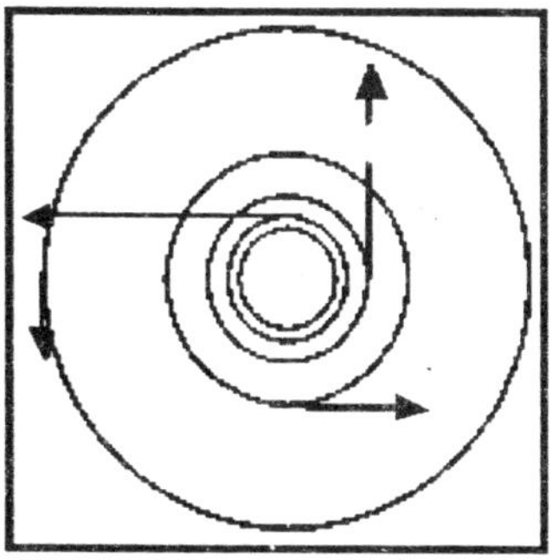

Fig. Velocity field of a vortex.

In cylindrical coordinates (r, θ, z) there is only one component of the velocity field, vθ. In calculating the circulation, the line element dI = rdθ $\hat{\theta}$, so that $v \cdot dI = v\theta r d\theta$. If the circulation is independent of the integration path, then we must have vθ = C/r, with *C* a constant.

The circulation is then,

$$\Gamma = \int_0^{2\pi} \left(\frac{C}{r} \right) r d\theta = 2\pi C,$$

so that C = G/2p. Therefore, the velocity field of a vortex is,

$$v(r) = \frac{\Gamma}{2\pi r} \hat{\theta} \text{ (vortex)}$$

By making the substitutions suggested above, we can use this result to find the magnetic field produced by a wire carrying a current *I*:

$$B(r) = \frac{\mu_o I}{2\pi r} \hat{\theta}$$

(current — carrying wire).

STREAM FUNCTION AND VELOCITY POTENTIAL

REVIEW OF VELOCITY-POTENTIAL CONCEPTS

This chapter presents examples of problems and their solutions for which the assumption of potential flow is appropriate. For low speed flows where viscous effects are neglected, the flow is irrotational and,

$$\tilde{N} \times V = 0$$

$$V \times \tilde{N} = f$$

$$u = \frac{\partial \phi}{\partial x}$$

$$y = \frac{\partial \phi}{\partial y}$$

$$w = \frac{\partial \phi}{\partial z}$$

The continuity equation, $\nabla V = 0$, now reduces to,

$$\nabla^2 V \frac{\partial^2 \phi}{\partial x^2} + \frac{\partial^2 \phi}{\partial y^2} + \frac{\partial^2 \phi}{\partial z^2} = 0$$

The momentum equation reduces to Bernoulli's equation:

$$\frac{\partial \phi}{\partial t} + \frac{P}{\rho} + \frac{1}{2} V^2 + gz = \text{const.}$$

REVIEW OF STREAM FUNCTION CONCEPTS

For plane incompressible flow in x-y coordinates, a stream function exists such that,

$$u \frac{\partial \Psi}{\partial y} \text{ and } v = -\frac{\partial \Psi}{\partial y}$$

The condition of irrotationality reduces to Laplace's equation for Y with,

$$\frac{\partial^2 \Psi}{\partial x^2} + \frac{\partial^2 \Psi}{\partial y^2} = 0$$

and for a solid surface - Ψ_{solid} = const.

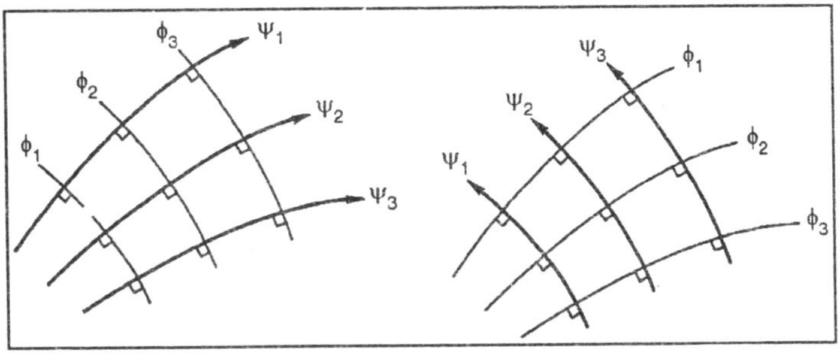

Fig. Streamlines and Potential Lines: (a) Typical Inviscid-flow Pattern: (b) Same as (a) with Roles Reversed

PLANE POLAR COORDINATES

Equations for plane polar velocity components are given below in term of polar coordinates (r, q) and the polar coordinate velocity potential, f, and stream function, Ψ.

$$v_r = \frac{\partial \phi}{\partial r} = \frac{1}{r}\frac{\partial \Psi}{\partial \theta}$$

$$v_\theta = \frac{1}{r}\frac{\partial \phi}{\partial \theta} = -\frac{\partial \Psi}{\partial r}$$

Laplace's equation now has the form in polar coordinates

$$\frac{1}{r}\frac{\partial}{\partial r}\left(r\frac{\partial \phi}{\partial r}\right) + \frac{1}{r^2}\frac{\partial^2 \phi}{\partial \theta^2} = 0$$

ELEMENTARY PLANE-FLOW SOLUTIONS

Three plane-flow solutions that are very useful in developing more complex potential flow solutions are:

Uniform stream, iU, in the x direction:

$$\Psi = U\,y \quad \phi = U\,x$$

Line Source or Sink

$$Y = m\,q \quad q = m \ln r$$

Line Vortex

$$\psi = -K \ln r \quad \phi = K\theta$$

In these expressions, the source strength, 'm', and vortex strength, 'K', have the dimensions of velocity times length, or $[L^2/t]$.

If the uniform stream is written in plane polar coordinates, we have,

Uniform stream, iU:

$$\Psi = U\, r\sin\theta \quad \phi = U\, r\cos\theta$$

For a uniform stream moving at an angle, a, relative to the x-axis, we can write,

$$u = U\cos\alpha = \frac{\partial\Psi}{\partial y} = \frac{\partial\phi}{\partial x}$$

$$v = U\sin\alpha = \frac{\partial\Psi}{\partial x} = \frac{\partial\phi}{\partial y}$$

After integration, we obtain the following expressions for the stream function and velocity potential:

$$\psi = U\,(y\cos\alpha - x\sin\alpha)$$
$$\phi = U\,(x\cos\alpha - y\sin\alpha)$$

CIRCULATION

The concept of fluid circulation is very useful in the analysis of certain potential flows, in particular those useful in aerodynamics analyses. Consider Figure 8.3 shown below:

We define the circulation, Γ, as the counterclockwise line integral of the arc length, dS times the velocity component tangent to the closed curve, C, e.g.

$$\Gamma = \oint_c V\cos\alpha\, ds = \int_c V\cdot ds$$

$$\Gamma = \int_c (u\,dx + v\,dy + w\,dz)$$

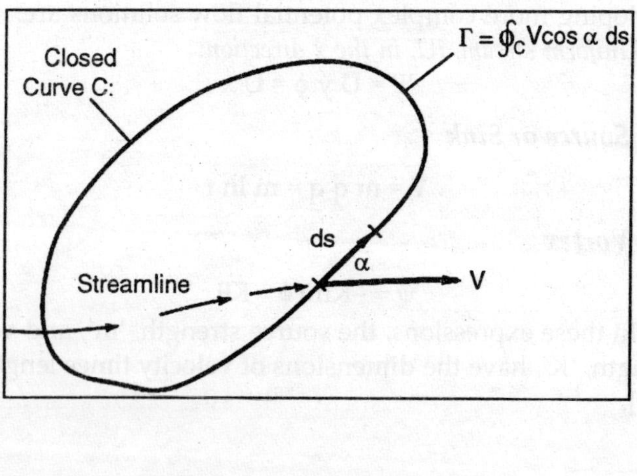

For most flows, this line integral around a closed path, starting and stopping at the same point, yields $\Gamma = 0$.

However,

For a vortex flow for which: $\phi = K\,\theta$

The integral yields: $\Gamma = 2\,\pi\,K$

An equivalent calculation can by made by defining a circular path of radius r around the vortex centre to yield,

$$\Gamma = \int_c v_\theta\, ds = \int_0^{2\pi} \frac{K}{r} r\, d\phi = 2\pi K$$

Key Point: A source or sink does not produce circulation. Without the presence of vortices, the circulation will be zero for any closed path around any number of sources or sinks.

SUPERPOSITION OF POTENTIAL FLOWS

Due to the mathematical character of the equations governing potential flows, the principle of superposition can be used to determine the solution for the flow which results from summing the velocity potential and stream functions two individual potential flow solutions.

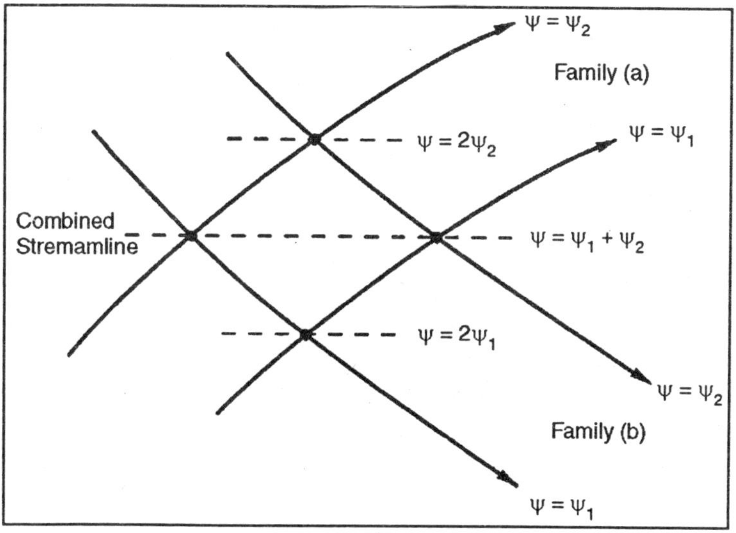

Fig. This is Shown Graphically.

Note that the value of the stream function at each

intersection equals the sum of the values of the stream functions crossing at the point. This would also be true for the velocity potential at any point in a combined flow.

Several classic examples of superposition of flows are presented as follows:

- Source m at (-a, 0) added to an equal sink at (+a, 0).

$$\psi = -m \tan^{-1} \frac{2ay}{x^2 + y^2 - a^2}$$

$$\phi = \frac{1}{2} m \ln \frac{(x+a)^2 + y^2}{(x-a)^2 + y^2}$$

The streamlines and potential lines are two families of orthogonal circles.

- Sink m plus a vortex K, both at the origin.

$\psi = m\theta - K \ln r \qquad \theta = m \ln r + K\theta$

The streamlines are logarithmic spirals swirling into the origin. They resemble a tornado or a bathtub vortex.

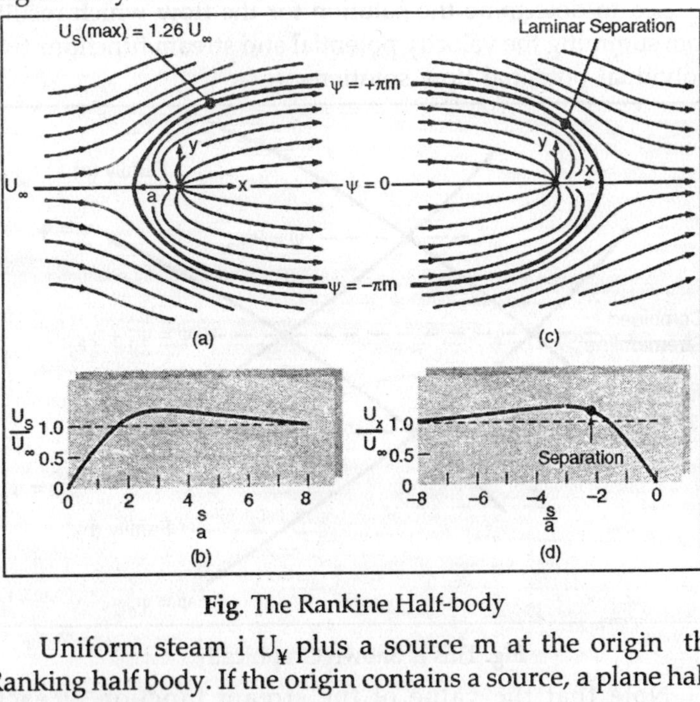

Fig. The Rankine Half-body

Uniform steam i $U_¥$ plus a source m at the origin the Ranking half body. If the origin contains a source, a plane half-body is formed with its nose to the left as shown below. If the origin contains a sink, $m < 0$, the half-body nose is to the right.. For both cases, the stagnation point is at a position $a = m / U_{¥}$ away from the origin.

- A uniform stream plus a source yields a half-body with stagnation point at $x = -a = -m/U_{¥}$.
- A uniform stream plus a sink with a stagnation point at $x = a = m/U_{¥}$. (b & d) variation of free stream velocity and thus free stream pressure with location.

Example: An offshore power plant cooling water intake has a flow rate of 1500 ft^3/s in water 30 ft deep as in Fig. If the tidal velocity approaching the intake is 0.7 ft/s,

(a) How far downstream does the intake effect extend

(b) How much width of tidal flow is entrained into the intake?

The sink strength is related to the volume flow, Q and water depth by,

$$m = \frac{Q}{2\pi b} = \frac{1500\,\text{ft}^3/\text{s}}{2\pi\,30\text{ft}} = 7.96\text{ft}^2/\text{s}$$

The length, a, of the downstream effect:

$$a = \frac{m}{U_\infty} = \frac{7.96\text{ft}^2/\text{s}}{0.7\,\text{ft/s}} = 11.4\,\text{ft}$$

The width, L, of flow entrained into the intake:

L = 2 p a = 2 p 11.4ft = 71 ft

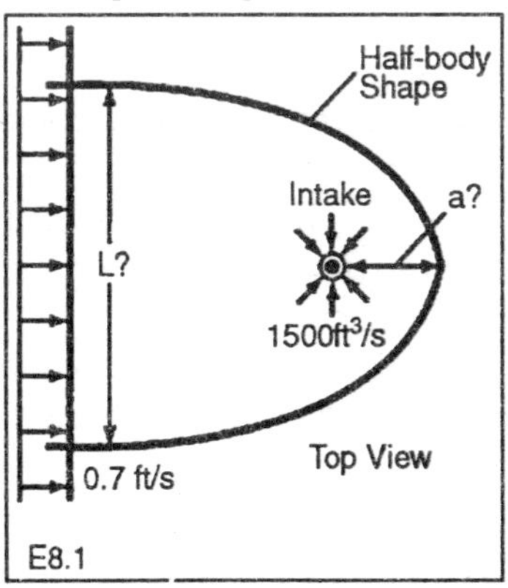

SOURCE

Source Flow speaks the language of modern, timeless design fluently. The taps of this line convey the clear, naturalness of water to the bathroom. Source Flow is both classic and unconventionally modern at the same time: reduced to the essential with the aura of poised composure, the taps give free reign to the wonderful power and freshness of the element of water.

Wonderfully natural and invigorating for body and spirit, the exhilarating element of water flows powerfully from the Source Flow cascade bath spout: Its energy can develop fully – a pure pleasure. A source of strength in everyday life that will make you yearn for tomorrow today with its timelessly-beautiful design. In three-dimensional flow, a point from which fluid issues at a uniform rate in all directions. In two-dimensional flow, a line normal to the planes of flow, from which fluid flows uniformly in all directions at right angles to the line.

SOURCE (SINK) FLOW

Potential function for 2D source of strength m at r = 0:

$$\phi=\frac{m}{2\pi}\text{lm}\, r$$

It satisfies $\nabla^2 \phi = 0$ (check Laplace's equation in polar coordinate in the keyword search utility), except at,

$$r=\sqrt{x^2+y^2}=0$$

(so must exclude r = 0 from flow)

- *Question*: Derive the equations for the velocity field for the 2D source.
  - *Hint*: expression for the gradient in polar coordinates (use the keyword utility: coordinate system - velocity vector)
- *Question*: Evaluate the outward volume flux.
  - *Hint*: Consider a contour which contains the 2D source.
  - *Hint*: Use Gauss theorem to deform the contour into a small circle of radius ε around the source.

- *Hint*: Evaluate the flux in the direction normal to the circle (radial velocity) and integrate along the circle.

If m < 0 ⇒ sink. Source with strength m located at (x_0, y_0):

$$\phi = \frac{m}{2\pi} \text{lm} \sqrt{(x-x_0)^2 + (y-y_0)^2}$$

$$\phi = \frac{m}{2\pi} \ln r \,(\text{Potential function}) \ll \psi = \frac{m}{2\pi} \theta \,(\text{Stream function})$$

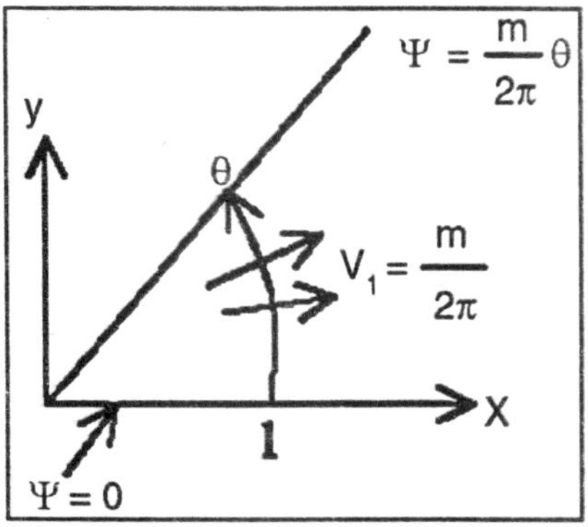

Fig. 3D source - Spherical Coordinates

A spherically symmetric solution:

$$\phi = \frac{1}{r}$$

(verify $\nabla^2 \phi = 0$ except at r = 0)

Define 3D source of strength m located at r = 0:

$$\phi = -\frac{m}{4\pi r}, \text{then } V_r = \frac{\partial \phi}{\partial r} = \frac{m}{4\pi r^1}, V_\theta, V_\varphi = 0$$

Chapter 3

Equations of Motion

OVERVIEW

We can use the Equations of Motion to calculate the speed of an object under different circumstances. These are quantities are involved in linear motion, movement in a straight line:

Quantity	Physics	CodeUnits
Distance	*s*	m
Speed at the start	*u*	m/s
Speed at the end	*v*	m/s
Acceleration	*a*	m/s^2
Time	*t*	s

- Speed at finish = speed at start + change in speed
  change in speed = acceleration × time.
  Speed at end = speed at start + (acceleration × time)
  Or in code:
  v = u + at
- $v^2 = u^2 + 2\ as$
- Distance = average speed × time
  Or in code:

$$s = \frac{(u+v)t}{2}$$

- $s = ut + \frac{1}{2}at^2$

***Maths Window*:**
Tips for Calculations

1. Write dwon a formula or equation $\quad s = ut + \frac{1}{2}at^2$

using standard symbols if possible

2. you may wish to write a "shoppling using" for the quantities in the formula /equation but this gets no marks.	s = 4.5m u = 0 v = not mentioned a = ? t = 0.90s
3. Substitute values into the formula/ equation without rearranging first (unless you know you won't make mistake).	$4.5 = 0 + \frac{1}{2}\alpha \times 0.90^2$ $4.5 = 0.405\alpha$
4. Rearrange and calculate your final answr	$a = 4.5 \div 0.405$
5. Write the answer using sensible significant figures and, ifnecessary, standar form. Add the proper units.	$a = 11.1,\ s^{-2}$

STREAMLINE AND INTEGRATION

The leading developer of Network on Chip (NoC) solutions andDuolog, a provider of SoC integration tools, today announced the integration of Duolog design tools with Arteris' NoC solution to provide designers a more streamlined and efficient way to integrate multiple semiconductor intellectual property (IP) blocks on a single system on ship (SoC) device.

The integration leverages Arteris' NoC solution for enabling high-performance on-chip interconnect and communications, and Duolog's Socrates™ Chip Integration Platform, which is a suite of tools for capturing, viewing and validating various elements of the infrastructure of complex SoCs.

The integration uses the industry standard IP-XACT format, a standard way to describe and handle IP from multiple sources. Using the Arteris NoCcompilerTM tool, SoC designers can quickly configure, instantiate, and connect NoC IP units to generate specific NoC instance(s), including full RTL and verification infrastructures. As part of this NoC IP instance generation, NoCcompiler generates an IP-XACT description which includes high-level interfaces, ports and memory map data.

The Duolog system uses the IP-XACT description to allow designers to capture the software view of the system from low-level IP registers and bitfields to the full system-level memory map. Its Socrates Bitwise™ register management tool not only captures the complete memory map infrastructure but provides real-time views from any point on the memory map. It generates a wide range of collateral including documentation, hardware design and verification infrastructures and software API models.

"With this integration, we have been able to add a new register to an IP to automatically generate a new software API and corresponding documentation at the chip level, all in less than a minute. For this level of turnaround time we need to ensure we clearly understand the NoC memory mapping, which is what our close collaboration with Arteris has allowed," said Ray Bulger, CEO of Duolog. "We have worked closely with Arteris and several of our customers to ensure that we align on the memory map content via the IP-XACT standard and have proved that this level of tool coupling can significantly improve the efficiency of streamlining our mutual customers' IP integration flow."

The streamlined flow also incorporates Duolog's Socrates Weaver™, a rules-based assembly engine that enables rapid and robust interface and port-level connectivity to ensure that the chip integration can happen sooner and more predictably.

"The integration of our NoC generation system with Duolog's viewing and validation capabilities provides a robust and efficient way to integrate complex SoCs,", said Charlie Janac, CEO of Arteris "Our customers can use the Arteris NoC IP and tools to rapidly generate NoC instance IPs to very quickly achieve their system performance goals. By combining our NoC approach with Duolog's tools, customers can integrate the Arteris NoC Instance IPs with the rest of the SoC to produce earlier software and hardware deliveries. By using standards such as IP-XACT to package the NoC IPs, it is an extremely interoperable approach that allows the use of a wide a variety of IP."

The integrated Arteris/Duolog solution will be demonstrated at the 2009 Design Automation Conference in San Francisco (July 26-31) in the Duolog Booth #2028.

About Duolog Technologies Ltd Duolog Technologies, the Collaborative Design Automation™ Company, is an award-winning developer of groundbreaking EDA tools that enable the flawless and rapid integration of today's increasingly complex SoC, ASIC and FPGA designs.

Duolog's Socrates chip integration platform employs a modular and extensible suite of tools for IO layer definition, IP packaging, connectivity and register management. The Socrates tools, built on the Eclipse platform and supporting the IP-XACT standard, shorten design cycles, reduce costs and greatly improve design quality through their Perfect-By-ConstructionTM methodology.

About Arteris Arteris, Inc. provides Network-on-Chip (NoC) interconnect IP, NoC Generation andverification tools to improve performance of system-on-chip (SoC) architectures for multimedia, mobile, telecom, and other applications. Arteris' NoC solution allows chip developers to implement scalable, efficient and high-performance SoC designs, overcoming limitations of traditional layered or pipelined bus-based architectures.

Results obtained by using Arteris NoC Solution include lower power, higher performance, lower risk of development and faster delivery of complex SoCs while increasing profits. Founded by networking experts, Arteris operates globally with headquarters in San Jose, California and an engineering centre in Paris, France. Arteris is a private company backed by a group of international investors including TVM Capital, Crescendo Ventures, Ventech, Synopsys and DoCoMo Capital.

About IP-XACT IP-XACT, a standard progressed by the SPIRIT consortium, defines a standard way for describing and handling multi-sourced IP components. This can enable automated design integration and configuration within multi-vendor design flows. The standardization effort in the SPIRIT consortium is driven by a coalition of leading semiconductor and EDA companies including Texas Instruments, ST, LSI, Freescale, NXP, Infineon, Cisco, Cadence, Synopsys and Mentor Graphics. Duolog is a contributing member of the Spirit consortium

ADDRESSING THE HYBRID MEDICAL RECORD

Today leading hospitals depend on electronic health record (EHR) solutions for access to critical patient information. Yet, these same organizations continue to rely on paper documents, such as paper chart records, consultation forms, progress notes, and other clinical documentation.

Streamline Health's accessANYware™ document management solutions bring these documents online and seamlessly integrate them with the electronic patient record. This gives clinicians access to more complete medical records, anytime, anywhere from within their clinical applications, saving time, enhancing patient care, and improving regulatory compliance.

To deliver a flexible, integrated solution, Streamline Health has developed its Streamline Health Integration Tools solution to enable accessANYware to support clinical information system interfaces.

IMPROVING WORKFLOW

Hospital benefit from a comprehensive solution that manages documents from within all of their enterprise software solutions and that streamlines and improves the efficiency of workflows. This Streamline Health integrated solution leverages technology to enable better patient care through:

- *A more complete patient record*: To gain the full picture of each patient's medical history that is critical to effective treatment, hospitals and clinics need patient records that include information from their patient care system.
- *Integration with existing applications*: Healthcare

systems have heterogeneous IT environments that include clinical systems as well as financial systems for revenue cycle management.

- *Automated workflows*: Healthcare organizations have numerous workflows that require them to route patient documents. These workflows include completion of the chart by physicians, coding, release of information, and tracking charts.
- *HIPAA compliance*: By providing access to documents in accordance with HIPAA regulations for privacy and security and that can be prepared to comply with additional regulations.
- *Scalability and reliability*: The system must also scale to support additional documents and users, and be reliable so as to support critical patient care on a 24/7 basis.

ONLINE HEALTH RECORDS

By *integrating Streamline Health's accessANYware and workflow solutions with their current clinical, financial and administrative applications, hospitals can achieve the following benefits*:

- Save physicians' time through remote chart access to pertinent patient information.
- Improve patient care decisions with ease of access to comprehensive health information.
- Enable teams of physicians and other providers to access charts simultaneously from multiple locations to improve team care.
- Allow coders and clinicians to work remotely via Web technology.
- Increase staff productivity by eliminating time consuming filing of charts.
- Decrease chart delinquency rates and suspension struggles.
- Empower users to make care-focused, cost-effective decisions.
- Comply with regulatory requirements by speeding

chart completion, and by controlling access to documents through strict role-based security

BERNOULLI EQUATION AND APPLICATIONS

A statement of the conservation of energy in a form useful for solving problems involving fluids. For a non-viscous, incompressible fluid in steady flow, the sum of pressure, potential and kinetic energies per unit volume is constant at any point

A special form of the Euler's equation derived along a fluid flow streamline is often called the Bernoulli Equation

$$\frac{\partial}{\partial s}\left(\frac{v}{2}+\frac{P}{\rho}+g\cdot h\right)=0 \quad (1)$$

where

v = flow speed

p = flow speed

ρ = density

g = gravity

h = height

$$\frac{v^2}{2}+\frac{P}{\rho}+g\cdot xh=(\text{Cons tan t}) \quad (2)$$

$$\frac{v^2}{2\cdot g}+\frac{P}{\gamma}+h=(\text{Cons tan t}) \quad (3)$$

where

$\gamma=\rho\cdot g$

$$\frac{\rho\cdot v^2}{2}+p=\text{cons tan t} \quad (4)$$

$$\frac{\rho\cdot v^2}{2}+pd \quad (5)$$

$$\frac{\rho\cdot v_1^2}{2}+p_1=\frac{\rho\cdot v_2^2}{2}+p_2=\text{Cons tan t} \quad (6)$$

For steady state incompressible flow the Euler equation becomes. If we integrate along the streamline it becomes. Can further be modified to by dividing by gravity.

HEAD OF FLOW

Equation (3) is often referred to the head because all elements has the unit of length.

Dynamic Pressure

(2) and (3) are two forms of the Bernoulli Equation for steady state incompressible flow. If we assume that the gravitational body force is negligible, (3) can be written as (4). Both elements in the equation have the unit of pressure and it's common to refer the flow velocity component as the dynamic pressure of the fluid flow (5).

Since energy is conserved along the streamline, (4) can be expressed as (6). Using the equation we see that increasing the velocity of the flow will reduce the pressure, decreasing the velocity will increase the pressure.

This phenomena can be observed in a venturi meter where the pressure is reduced in the constriction area and regained after. It can also be observed in a pitot tube where the stagnation pressure is measured. The stagnation pressure is where the velocity component is zero.

Example

Bernoulli Equation and Flow from a Tank through a small Orifice. Liquid flows from a tank through a orifice close to the bottom. The Bernoulli equation can be adapted to a streamline from the surface to the orifice as (e1):

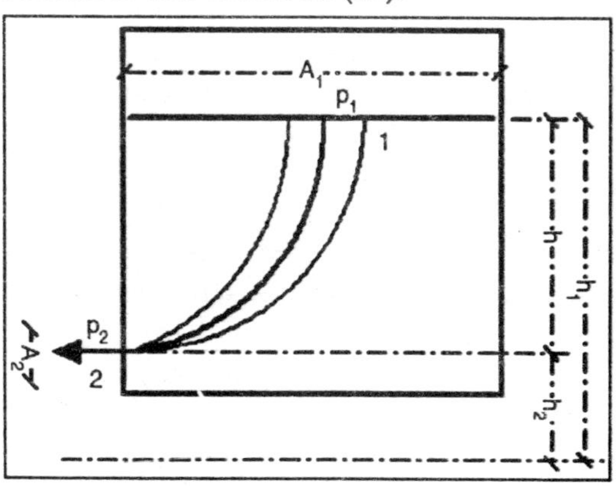

$$g \cdot h_1 + \frac{p_1}{\rho} + \frac{V_1^2}{2} = g \cdot h_2 + \frac{p_2}{\rho} + \frac{V_2^2}{2} \quad \text{(ei)}$$

where,

$$h = h_1 - h_2 \quad \text{(e2)}$$

$$V_1 = \left(\frac{A_2}{A_1}\right) \cdot V_2 \quad \text{(e3)}$$

$$V_2 = \sqrt{\frac{2}{\left(1 - \frac{A_2^2}{A_1^2}\right)}} \cdot \left(\frac{p_1 - p_2}{\rho} + g \cdot h\right) \quad \text{(e4)}$$

$$V_2 = \sqrt{2 \cdot g \cdot h} \quad \text{(e5)}$$

$$V_2 = c\sqrt{2 \cdot g \cdot h} \quad \text{(e5b)}$$

$$V_2 = \sqrt{\left(\frac{2 \cdot (p_1 - p_2)}{\rho}\right)}$$

Since (1) and (2)'s heights from a common reference is related as (e2), and the equation of continuity can be expressed as (e3), it's possible to transform (e1) to (e4).

Vented tank

A special case of interest for equation (e4) is when the orifice area is much lesser than the surface area and when the pressure inside and outside the tank is the same - when the tank has an open surface or "vented" to the atmosphere. At this situation the (e4) can be transformed to (e5).

"The velocity out from the tank is equal to speed of a freely body falling the distance *h*." - also known as Torricelli's Theorem.

Example

Outlet velocity from a vented tank,

The outlet velocity of a tank with height *10 m* can be calculated as,

$$V_2 = (2\ (9.81\ m/s^2)\ (10\ m))^{1/2}$$
$$= 14\ m/s$$

Pressurized Tank

If the tanks is pressurized so that product of gravity and height (g h) is much lesser than the pressure difference divided by the density, (e4) can be transformed to (e6).

The velocity out from the tank depends mostly on the pressure difference.

Example

Outlet velocity from a pressurized tank.

The outlet velocity of a pressurized tank where,

$p_1 = 0.2\ MN/m^2$

$p_2 = 0.1\ MN/m^2$

$A_2/A_1 = 0.01$

$h = 10\ m$

can be calculated as:

$V_2 = ((2/(1-(0.01)^2)\ ((0.2\ 10^6\ N/m^2) - (0.1\ 10^6\ N/m^2))/(1000\ kg/m^3) + (9.81\ m/s^2)(10\ m)))^{1/2} = 19.9\ m/s$

A Pitot-static tube can measure the fluid flow velocity by converting the kinetic energy in the fluid flow into potential energy.

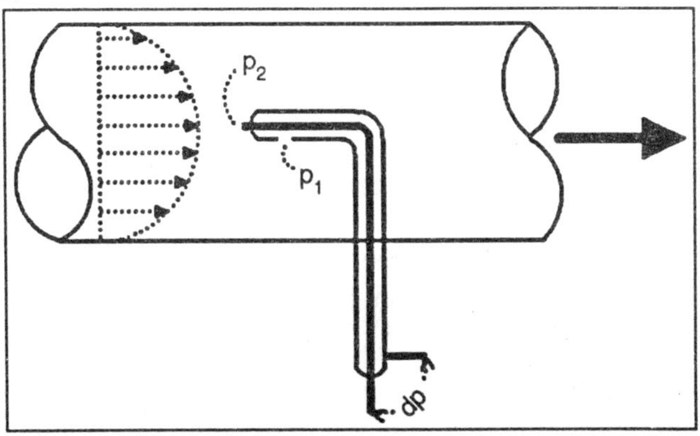

The principle is based on the Bernoulli Equation where each term can be interpreted as a form of pressure

$p + 1/2\ \rho\ v^2 + Yh$ = constant along a streamline (1)

where,

p = Static pressure (relative to the moving fluid) (Pa)

ρ = Density (kg/m^3)

Y = Specific weight (kN/m^3)
v = Flow velocity (m/s)
g = Acceleration of gravity (m/s^2)
h = Elevation height (m)

Each term of this equation has the dimension force per unit area - N/m^2 or in imperial units *psi*, lb/ft^2.

Static Pressure

The first term - *p* - is the static pressure. It is static relative to the moving fluid and can be measured through an flat opening in parallel to the flow.

Dynamic Pressure

The second term - $1/2\ \rho\ v^2$ - is called the dynamic pressure.

Hydrostatic Pressure

The third term - Y h - is called the hydrostatic pressure. It represent the pressure due to change in elevation.

Stagnation Pressure

Since the Bernoulli Equation states that the energy along the streamline is constant, (1) can be modified to

$$p_1 + 1/2\ \rho\ v_1^2 + Y\ h_1$$
$$= p_2 + 1/2\ \rho\ v_2^2 + Y\ h_2$$
$$= \text{constant along the streamline} \quad (2)$$

where,

suffix $_1$ is a point the free flow upstream

suffix $_2$ is the stagnation point where the velocity in the flow is zero

Flow Velocity

In a measuring point we regard the hydrostatic pressure as a constant, $h_1 = h_2$ and this part can be eliminated. Since v_2 is zero, (2) can be modified to:

$$p_1 + 1/2\ r\ v_1^2 = p_2 \quad (3)$$

or,

$$v_1 = [2\ (p_2 - p_1)\ /\ r]^{1/2} \quad (4)$$

where,

$p_2 - p_1 = dp$ (differential pressure)

With (4) it's possible to calculate the flow velocity in point 1 - the free flow upstream - if we know the differential pressure difference $dp = p_2 - p_1$ and the density of the fluid.

PITOT TUBE

The pitot tube is a simple and convenient instrument to measure the difference between static, dynamic and total pressure (or head).

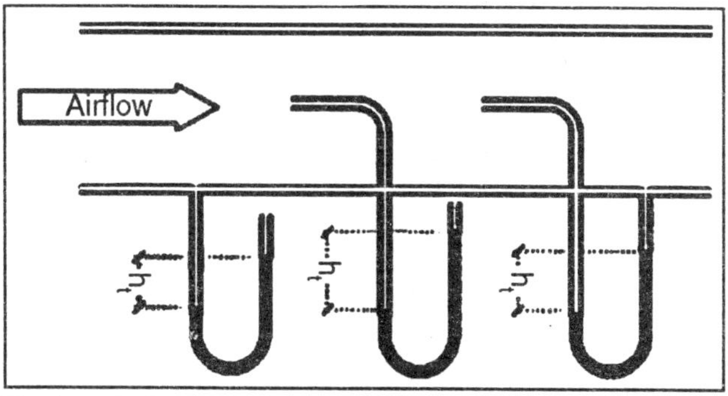

The head - *h* - (or pressure difference - *dp*) can be measured and calculated with u-tube manometers, electronic pressure transmitters or similar instrumentation.

The Bernoulli equation can be applied to a great many situations not just the pipe flow we have been considering up to now. In the following sections we will see some examples of its application to flow measurement from tanks, within pipes as well as in open channels.

If a stream of uniform velocity flows into a blunt body, the stream lines take a pattern similar to this:

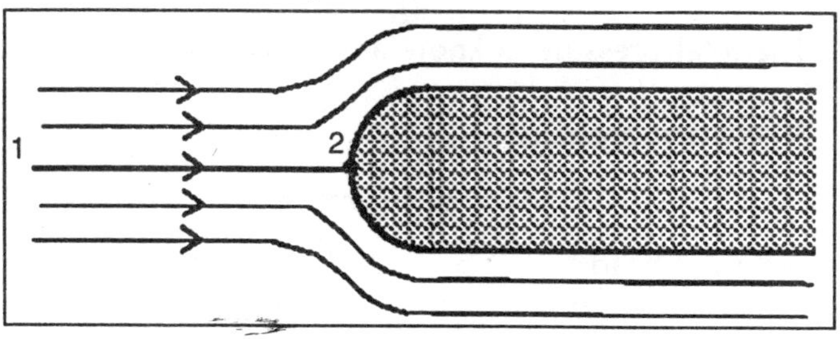

Fig. Streamlines Around a Blunt Body

Note how some move to the left and some to the right. But one, in the centre, goes to the tip of the blunt body and stops. It stops because at this point the velocity is zero - the fluid does not move at this one point. This point is known as the *stagnation point.*

From the Bernoulli equation we can calculate the pressure at this point. Apply Bernoulli along the central streamline from a point upstream where the velocity is u_1 and the pressure p_1 to the stagnation point of the blunt body where the velocity is zero, $u_2 = 0$. Also $z_1 = z_2$.

$$\frac{p_1}{\rho g}+\frac{u_1^2}{2g}+z_1=\frac{p_2}{\rho g}+\frac{u_2^2}{2g}+z_2$$

$$\frac{p_1}{\rho}+\frac{u_1^2}{2}=\frac{p_2}{\rho}$$

$$p_2=p_1+\frac{1}{2}\rho u_1^2$$

This increase in pressure which bring the fluid to rest is called the *dynamic pressure.*

Dynamic pressure =

$$\frac{1}{2}\rho u_1^2$$

or converting this to head (using,

$$h=\frac{p}{\rho g}).$$

Dynamic head =

$$\frac{1}{2g}u_1^2.$$

The total pressure is know as the *stagnation pressure* (or *total pressure*).

Stagnation pressure =

$$p_1+\frac{1}{2}\rho u_1^2$$

or in terms of head.

Stagnation head =

$$\frac{p_1}{\rho g}+\frac{1}{2g}u_1^2$$

The blunt body stopping the fluid does not have to be a solid. I could be a static column of fluid. Two piezometers, one as normal and one as a Pitot tube within the pipe can be used in an arrangement shown below to measure velocity of flow.

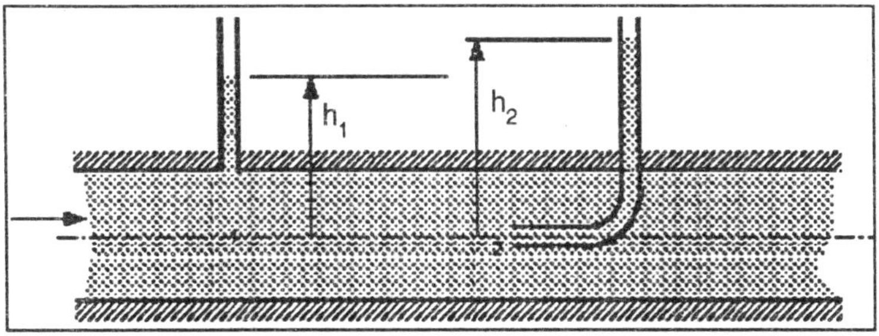

Fig. A Piezometer and a Pitot Tube

Using the above theory, we have the equation for p_2,

$$p_2 = p_1 + \frac{1}{2}\rho u_1^2$$

$$\rho g h_1 = \rho g h_1 + \frac{1}{2} r u_1^2$$

$$u = \sqrt{2g(h_2 - h_1)}$$

We now have an expression for velocity obtained from two pressure measurements and the application of the Bernoulli equation.

Pitot Static Tube

The necessity of two piezometers and thus two readings make this arrangement a little awkward. Connecting the piezometers to a manometer would simplify things but there are still two tubes.

The *Pitot static* tube combines the tubes and they can then be easily connected to a manometer. A Pitot static tube is shown below.

The holes on the side of the tube connect to one side of a manometer and register the *static head, (h_1)*, while the central hole is connected to the other side of the manometer to register, as before, the *stagnation head (h_2)*.

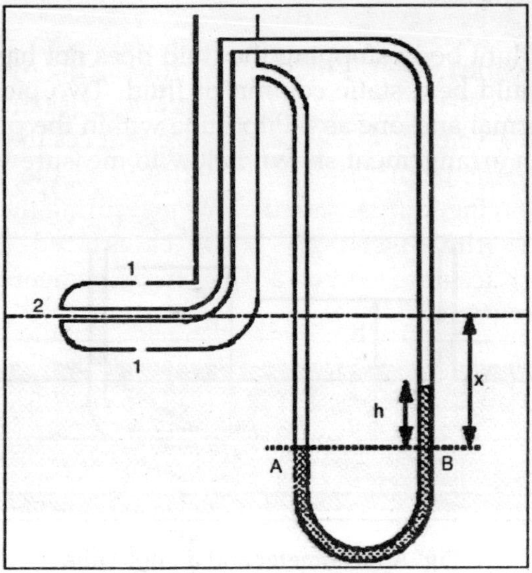

Fig. A Pitot-static tube

Consider the pressures on the level of the centre line of the Pitot tube and using the theory of the manometer:

$P_A = P_2 + pgX$

$P_B = P_1 + pg\ (X - h) + \rho_{max}gh$

$P_A = P_B$

$P_2 + \rho gX = p_1 + \rho g(X - h) + \rho_{max}gh$

We know that $P_2 = P_{static} =$

$P_1 + \frac{1}{2}\rho u_1^2,$

substituting this in to the above gives,

$$P_1 + hg(\rho_{max} - \rho) = P_1 + \frac{\rho u_1^2}{2}$$

$$u_1 = \sqrt{\frac{2gh(\rho_m - \rho)}{\rho}}$$

The Pitot/Pitot-static tubes give velocities at points in the flow. It does not give the overall discharge of the stream, which is often what is wanted. It also has the drawback that it is liable to block easily, particularly if there is significant debris in the flow.

Venturi Meter

The Venturi meter is a device for measuring discharge in a pipe. It consists of a rapidly converging section which increases the velocity of flow and hence reduces the pressure. It then returns to the original dimensions of the pipe by a gently diverging 'diffuser' section. By measuring the pressure differences the discharge can be calculated. This is a particularly accurate method of flow measurement as energy loss are very small.

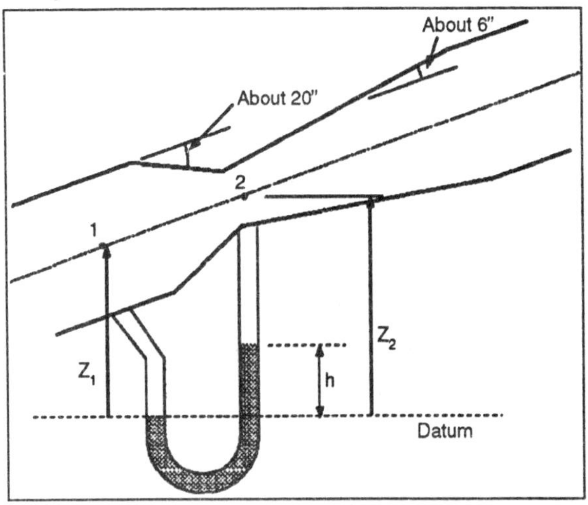

Fig. A Venturi Meter

Applying Bernoulli along the streamline from point 1 to point 2 in the narrow *throat* of the Venturi meter we have,

$$\frac{P_1}{\rho g}+\frac{u_1^2}{2g}+z_1=\frac{P_2}{\rho g}+\frac{u_2^2}{2g}+z_2$$

By the using the continuity equation we can eliminate the velocity u_2,

$$Q=u_1A_1=u_2A_2$$

$$u_2=\frac{u_1A_1}{A_2}$$

Substituting this into and rearranging the Bernoulli equation we get,

$$\frac{P_1 - P_2}{\rho g} + z_1 - z_2 = \frac{u_1^2}{2g}\left[\left(\frac{A_1}{A_2}\right)^2 - 1\right]$$

$$u_1 = \sqrt{\frac{2g\left[\frac{P_1 - P_2}{\rho g} + z_1 - z_2\right]}{\left(\frac{A_1}{A_2}\right)^2 - 1}}$$

To get the theoretical discharge this is multiplied by the area. To get the actual discharge taking in to account the losses due to friction, we include a coefficient of discharge,

$$Q_{ideal} = u_1 A_1$$

$$Q_{ideal} = C_d Q_{ideal} = C_d u_1 A_1$$

$$Q_{ideal} = C_d A_1 A_2 \sqrt{\frac{2g\left[\frac{P_1 - P_2}{\rho g} + z_1 - z_2\right]}{A_1^2 - A_2^2}}$$

This can also be expressed in terms of the manometer readings,

$$P_1 + rgz_1 = P_2 + r_{max}gh + rg(z_2 - h)$$

$$\frac{P_1 - P_2}{\rho g} + z_1 - z_2 = h\left(\frac{\rho_{max}}{\rho} - 1\right)$$

Thus the discharge can be expressed in terms of the manometer reading:

$$Q_{actual} = C_d A_1 A_2 \sqrt{\frac{2gh\left(\frac{\rho_{max}}{\rho} - 1\right)}{A_1^2 - A_2^2}}$$

Notice how this expression does not include any terms for the elevation or orientation (z_1 or z_2) of the Venturimeter. This means that the meter can be at any convenient angle to function.

The purpose of the diffuser in a Venturi meter is to assure gradual and steady deceleration after the throat. This is designed to ensure that the pressure rises again to something near to the original value before the Venturi meter.

The angle of the diffuser is usually between 6 and 8 degrees. Wider than this and the flow might separate from the walls resulting in increased friction and energy and pressure loss.

If the angle is less than this the meter becomes very long and pressure losses again become significant. The efficiency of the diffuser of increasing pressure back to the original is rarely greater than 80%.

Flow Through A Small Orifice

We are to consider the flow from a tank through a hole in the side close to the base. The general arrangement and a close up of the hole and streamlines are shown in the figure below

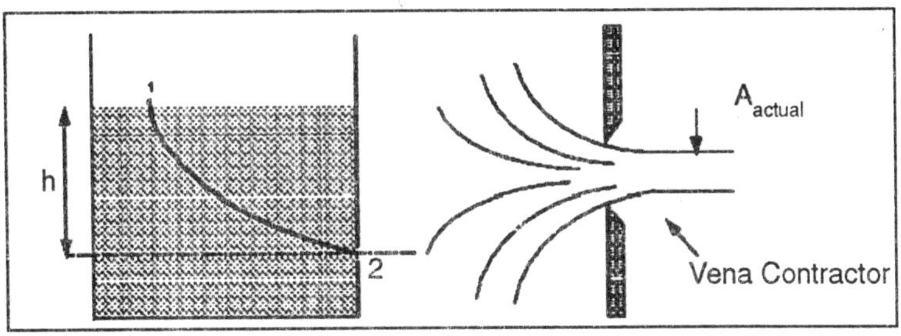

Fig.Tanket and Streamlines of
Flow out of the Sharp Edged Orifice

The shape of the holes edges are as they are (sharp) to minimise frictional losses by minimising the contact between the hole and the liquid - the only contact is the very edge.

Looking at the streamlines you can see how they contract after the orifice to a minimum value when they all become parallel, at this point, the velocity and pressure are uniform across the jet.

This convergence is called the *vena contracta.* (From the Latin 'contracted vein'). It is necessary to know the amount of contraction to allow us to calculate the flow.

We can predict the velocity at the orifice using the Bernoulli equation. Apply it along the streamline joining point

1 on the surface to point 2 at the centre of the orifice. At the surface velocity is negligible ($u_1 = 0$) and the pressure atmospheric ($p_1 = 0$).At the orifice the jet is open to the air so again the pressure is atmospheric ($p_2 = 0$). If we take the datum line through the orifice then $z_1 = h$ and $z_2 = 0$, leaving

$$h = \frac{u_2^2}{2g}$$

$$u_2 = \sqrt{2gh}$$

This is the theoretical value of velocity. Unfortunately it will be an over estimate of the real velocity because friction losses have not been taken into account. To incorporate friction we use the coefficient of velocity to correct the theoretical velocity,

$$u_{actual} = C_v u_{theoretical}$$

Each orifice has its own coefficient of velocity, they usually lie in the range(0.97 – 0.99).

To calculate the discharge through the orifice we multiply the area of the jet by the velocity. The actual area of the jet is the area of the vena contracta not the area of the orifice. We obtain this area by using a coefficient of contraction for the orifice

$$A_{actual} = C_c A_{orifice}$$

So the discharge through the orifice is given by

$$Qactual = A_{actua} u_{actual}$$

$$= C_c C_v \, A_{orifice} u_{theoretical}$$

$$= C_d A_{orifice} \sqrt{2gh}$$

Where C_d is the coefficient of discharge, and $C_d = C_c \, C_v$

Time for a Tank to Empty

We now have an expression for the discharge out of a tank based on the height of water above the orifice. It would be useful to know how long it would take for the tank to empty.

As the tank empties, so the level of water will fall. We can get an expression for the time it takes to fall by integrating the expression for flow between the initial and final levels.

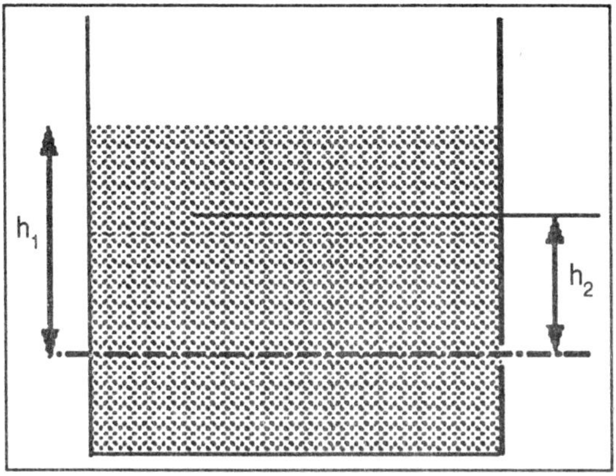

Fig. Tank Emptying from Level h_1 to h_2.

The tank has a cross sectional area of *A*. In a time *dt* the level falls by *dh* or the flow out of the tank is,

$$Q = Av$$

$$Q = -A\frac{\delta h}{\delta t}$$

(–ve sign as *dh* is falling), Rearranging and substituting the expression for *Q* through the orifice gives,

$$\delta h = \frac{-A}{C_d A_o \sqrt{2g}} \frac{\delta h}{\sqrt{h}}$$

This can be integrated between the initial level, h_1, and final level, h_2, to give an expression for the time it takes to fall this distance,

$$t = \frac{-A}{C_d A_o \sqrt{2g}} \int_{h_1}^{h_2} \frac{\delta h}{\sqrt{h}}$$

$$= \frac{-A}{C_d A_o \sqrt{2g}} \left[2\sqrt{h}\right]_{h_1}^{h_2}$$

$$= \frac{-2A}{C_d A_o \sqrt{2g}} \left[\sqrt{h_2} - \sqrt{h_1}\right]$$

Submerged Orifice: We have two tanks next to each other (or one tank separated by a dividing wall) and fluid is to flow between them through a submerged orifice. Although difficult to see, careful measurement of the flow indicates that the submerged jet flow behaves in a similar way to the jet in air in that it forms a vena contracta below the surface. To determine the velocity at the jet we first use the Bernoulli equation to give us the ideal velocity. Applying Bernoulli from point 1 on the surface of the deeper tank to point 2 at the centre of the orifice, gives,

$$\frac{P_1}{\rho g}+\frac{u_1^2}{2g}+z_1=\frac{p_2}{\rho g}+\frac{u_2^2}{2g}+z_1$$

$$0+0+h_1=\frac{\rho g h_2}{\rho g}+\frac{u_2^2}{2g}+0$$

$$u_2=\sqrt{2g(h_1-h_2)}$$

i.e. the ideal velocity of the jet through the submerged orifice depends on the*difference* in head across the orifice. And the discharge is given by

$$Q = C_d A_o u$$

$$=C_d A_o \sqrt{2g(h_1-h_2)}$$

Time for Equalisation of Levels in Two Tanks:

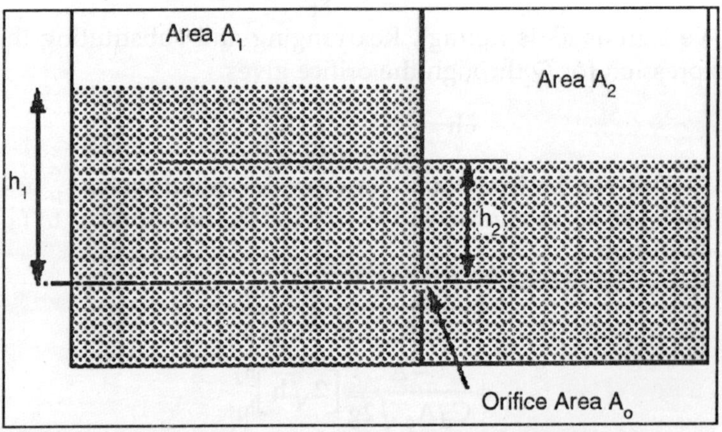

Fig. Two Tanks of Initially Different Levels Joined by an Orifice

By a similar analysis used to find the time for a level drop in a tank we can derive an expression for the change in levels when there is flow between two connected tanks.

Applying the continuity equation:

$$Q=-A_1\frac{\delta h_1}{\delta t}=A_2\frac{\delta h_2}{\delta t}$$

$$Q\delta t = -A_1\,\delta h_1 = A_2\,\delta h_2$$

Also we can write $-\delta h_1 + \delta h_2 = \delta h$,

So,

$$-A_1 dh_1 + A_2 dh_2 - A_2 dh$$

$$\delta h_1 = \frac{A_2 \delta h}{A_1 + A_2}$$

Then we get,

$$Q\,dt = -A_1 dh_1$$

$$C_d A_o \sqrt{2g(h_1 - h_2)}\,\delta t = \frac{A_1 A_2}{A_1 + A_2}\delta h$$

Re-arranging and integrating between the two levels we get,

$$\delta t = \frac{A_1 A_2}{(A_1 + A_2) C_d A_o \sqrt{2g}} \frac{\delta h}{h}$$

$$t = \frac{A_1 A_2}{(A_1 + A_2) C_d A_o \sqrt{2g}} \int_{h_{initial}}^{h_{final}} \frac{\delta h}{\sqrt{h}}$$

$$= \frac{2 A_1 A_2}{(A_1 + A_2) C_d A_o \sqrt{2g}} \left[\sqrt{h}\right]_{h_{initial}}^{h_{final}}$$

$$= \frac{2 A_1 A_2}{(A_1 + A_2) C_d A_o \sqrt{2g}} \left[\sqrt{h_{initial}} - \sqrt{h_{final}}\right]$$

(remember that h in this expression is the *difference* in height between the two levels $(h_2 - h_1)$ to get the time for the levels to equal use $h_{initial} = h_1$ and $h_{final} = 0$).

Thus we have an expression giving the time it will take for the two levels to equal.

Flow Over Notches and Weirs

A notch is an opening in the side of a tank or reservoir which extends above the surface of the liquid. It is usually a device for measuring discharge. A weir is a notch on a larger scale - usually found in rivers. It may be sharp crested but also may have a substantial width in the direction of flow - it is used as both a flow measuring device and a device to raise water levels.

Weir Assumptions

We will assume that the velocity of the fluid approaching the weir is small so that kinetic energy can be neglected. We will also assume that the velocity through any elemental strip depends only on the depth below the free surface. These are acceptable assumptions for tanks with notches or reservoirs with weirs, but for flows where the velocity approaching the weir is substantial the kinetic energy must be taken into account (e.g. a fast moving river).

A General Weir Equation

To determine an expression for the theoretical flow through a notch we will consider a horizontal strip of width *b* and depth *h* below the free surface, as shown in the figure below.

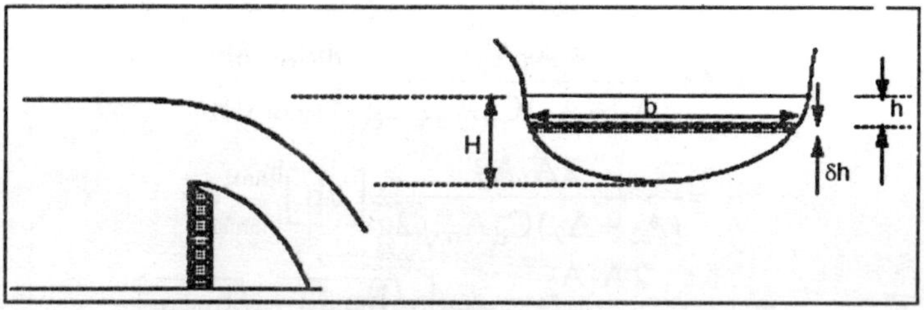

Fig. Elemental Strip of Flow Through a Notch

velocity through the strip, $u = \sqrt{2gh}$

discharge through the e strip, $\delta Q = Au = b\,\delta h\sqrt{2gh}$

integrating from the free surface, h = 0, to the weir crest, h = H gives the expression for the total theoretical discharge

$$Q_{theoretical} = \sqrt{2g}\int_{o}^{H} bh^{\frac{1}{2}}dh$$

This will be different for every differently shaped weir or notch. To make further use of this equation we need an expression relating the width of flow across the weir to the depth below the free surface.

Rectangular Weir

For a rectangular weir the width does not change with

depth so there is no relationship between *b* and depth *h*. We have the equation,

$$b = \text{constant} = B$$

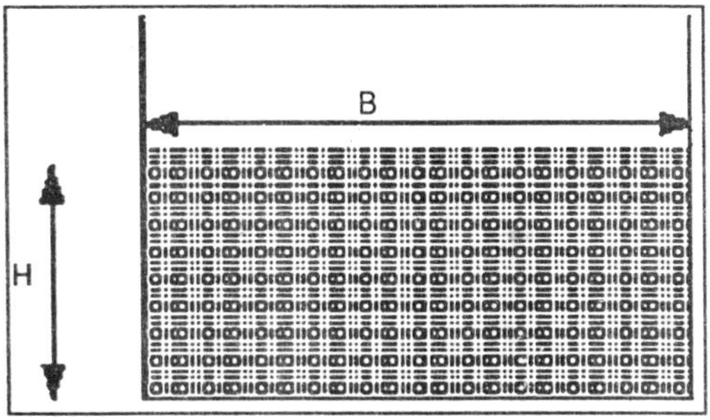

Fig. A Rectangular Weir

Substituting this into the general weir equation gives,

$$Q_{theoretical} = B\sqrt{2g}\int_{o}^{H} h\frac{1}{2}dh$$

$$= \frac{2}{3}B\sqrt{2g H^{3/2}}$$

To calculate the actual discharge we introduce a coefficient of discharge, C_d, which accounts for losses at the edges of the weir and contractions in the area of flow, giving,

$$Q_{actual} = C_d \frac{2}{3} B\sqrt{2g H^{3/2}}$$

'V' Notch Weir

For the "V" notch weir the relationship between width and depth is dependent on the angle of the "V".

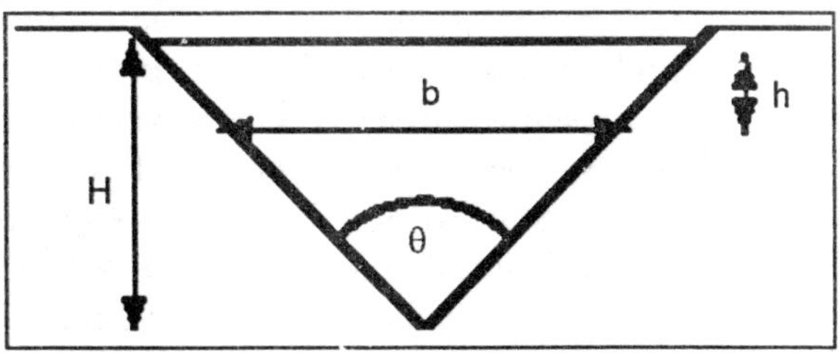

Fig. "V" Notch, or Triangular, Weir Geometry.

If the angle of the "V" is θ then the width, *b*, a depth *h* from the free surface is,

$$b=2(H-h)\tan\left(\frac{\theta}{2}\right)$$

So the discharge is:

$$Q_{theoretical}=2\sqrt{2g}\tan\left(\frac{\theta}{2}\right)\int_{o}^{H}(H-h)h^{1/2}\,dh$$

$$2\sqrt{2g}\tan\left(\frac{\theta}{2}\right)\left[\frac{2}{5}Hh^{3/2}-\frac{2}{5}h^{5/2}\right]_{o}^{H}$$

$$=\frac{8}{15}\sqrt{2g}\tan\left(\frac{\theta}{2}\right)H^{5/2}$$

And again, the actual discharge is obtained by introducing a coefficient of discharge,

$$Q_{actual}=C_d\frac{8}{15}\sqrt{2g}\tan\left(\frac{\theta}{2}\right)H^{5/2}$$

BEND AND VENTURI METER

INTRODUCTION

The notes on this page related to the methods of measuring flow using devices which are based on bernoulli's equation. There are many other devices which are convenient to use and are very accurate which are based on other principles including vortex shedding, ultrasonics (doppler), turbines, and variable orifice. To obtain information on these devices please consult the linked sites at the bottom of this page.

Pitot tubes, Orifice Plates, Nozzles and Venturi meters are established methods of measuring flows in pipelines. These methods relate the pressure difference across the upstream and downstream sides of the units to the pipeline fluid velocities. In modern piping systems various high technology methods many of which are non-intrusive are replacing these systems. Orifice plates and nozzles are also used as flow balancing and or limiting devices.

For information on flow resistance provided by an orifice plate refer Expansion/Contractions/Orifice Plates

Pitot tubes are specially designed probes inserted into pipes to establish the flow velocity at fixed points in the pipe bore. The flow rate is established using special techniques.

Orifice plates are low cost devices consisting of thin plates trapped between flanges. The orifice plate includes a sized hole with a downstream bevel, through which the fluid flows. The flanges include tapping points to measure the pressure upstream and downstream of the plates. The accuracy of the orifice plate method is about ± 2%

Nozzles are the same as orifice plates except that the thin plate is replaced by a contoured nozzle. The accuracy of the nozzle method is about ± 1%

The venturi is a converging length of pipe followed by a short parallel throat and then a divergence. The accuracy of the venturi method is better than ± 1%

The main differences in the devices are that the orifice plate results in signficant losses, the nozzle has relatively low losses and the venture meter and the pitot tubes are very efficient. The venturi meter is also the most accurate followed by the nozzle and the orifice plate. The orifice plate system is the most widely used because it is the cheapest and most convenient to install and maintain.

SYMBOLS

A = Area (m2)

A_2= Area of Orifice(m_2)

a = Speed of sound (m/s)

C_d = Coefficient of Discharge

C_c = Coefficient of Contraction

r = density (kg/m3)

r_1 = density at inlet condtions (kg/m3)

g = acceleration due to gravity (m/s2)

e = Expansion factor

h = fluid head (m)

L = Pipe length (m)

p_{gauge} - gauge pressure (N /m2)

p_{atm} - atmospheric pressure (N /m2

p_s= surface pressure (N /m^2)

Q = Volume flow rate (m^3/s)

Q_m = Mass flow rate = Qr (kg/s)

Re = Reynolds Number = u.r D/m

s = specific volume (m^3 /kg)

u = fluid velocity (m/s)

v = specific volume (m^3/kg)

v_1 = specific volume at inlet conditions(m^3/kg)

x = depth of centroid (m)

m = mass (kg)
$\underline{m}$ = mass flow rate (kg/s)
P_1 = Inlet fluid pressure (gauge) (N/m^2)
P_2 = Outlet fluid pressure
$\underline{P}_1$ = Inlet fluid pressure (abs) (N/m^2)
$\underline{P}_2$ = Outlet fluid pressure (abs) (N/m^2)
$\underline{P}$ - Absolute pressure (N/m^2)

θ = slope (radians)
β = Ratio of largest pipe dia to small diameter
τ = shear stress (N/m^2)
μ = dynamic viscosity (Pa.s)
ν = kinematic viscosity (m^2/s–^{1})
υ = Specific volume (m^3/kg)
γ = Ratio of Specific Heats

RELEVANT STANDARDS

The following standards provide detailed information on measuring fluid flow using venturis, orifice plates and nozzles.

Moff = Measurement of fluid flow by means of pressure differential devices inserted in circular cross-section conduits running full.

- *BS EN ISO* 5167-1:*2003 Moff*: General principles and requirements
- *BS EN ISO* 5167-2:2003 Moff: Orifice plates
- *BS EN ISO* 5167-3:2003 Moff: Nozzles and Venturi nozzles
- *BS EN ISO* 5167-4:2003 Moff: Venturi tubes

Note: BS 1042:Part 1(5 sections) and Part 3:Standards withdrawn and replaced by above standards.

- MoFFiCC = Measurement of Fluid Flow in closed conduits
- BS 1042-1.4:1992 BS 1042-2.1:1983, ISO 3966:1977 MoFFi CC. Velocity area methods. Method using Pitot static tubes
- BS 1042-2.2:1983, ISO 7145:1982 MoFFiCC. Velocity area methods. Method of measurement of velocity at one point of a conduit of circular cross section
- BS 1042-2.3:1984, ISO 7194:1983 MoFFiCC. Velocity area methods. Methods of flow measurement in swirling or asymmetric flow conditions in circular ducts by means of current-meters or Pitot static tubes

- BS 1042-2.4:1989, ISO 3354:1988 MoFFiCC. Velocity area methods. Method of measurement of clean water flow using current meters in full conduits and under regular flow conditions
- BS 1042:Part 1:Section 1.5:1997 MoFFiCC. Pressure differential devices. Guide to the effect of departure from the conditions specified in BS EN ISO 5167-1
- BS 1042:Part 1:Section 1.5:1987 MoFFiCC. Pressure differential devices. Guide to the effect of departure from the conditions specified in Section 1.1

PITOT METER

Consider three glass tubes positioned in a pipe which is carrying flowing fluid,

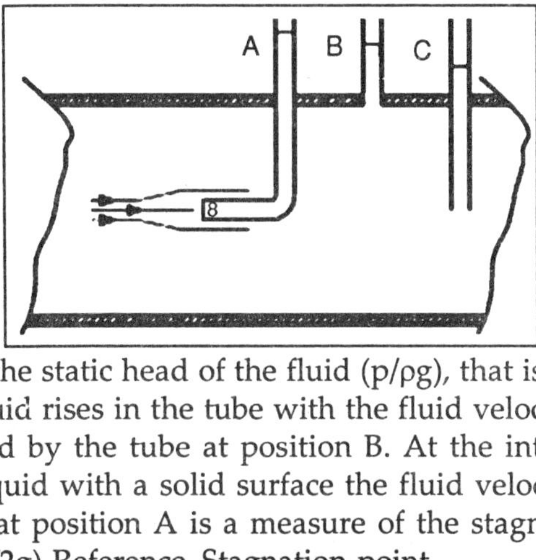

Now the static head of the fluid ($p/\rho g$), that is the height that the fluid rises in the tube with the fluid velocity at zero, is indicated by the tube at position B. At the interface of a flowing liquid with a solid surface the fluid velocity is zero. The head at position A is a measure of the stagnation head ($p/\rho g + u^2/2g$) Reference..Stagnation point.

Now if the flowing fluid was in an open stream that static pressure would be the atmospheric pressure and the stagnation head would be simply the level of fluid in the manometer above the surface level of the flowing fluid. For an enclosed stream the velocity head is the difference in level of the static head as measured by tube B and the stagnation head as indicated by Tube A. The level of fluid in the tube at C is not useful because the fluid is flowing past the end of the tube.

The figure below shows a typical design of pitot tube flow meter. The inner tube pressure is the stagnation pressure and the annulus surrounding the inner tube is at the static pressure i.e. it indicates the pressure at the surface of the pitot tube which is static. If ρ_m is the density of the manometer fluid and ρ is the density of the flowing fluid then the the fluid velocity results from the equation.

$u = c\ \Phi\ (2\ \Delta p/\rho)\ \Delta p = (\rho\ m-\rho)gx$....and.... $\Delta h = [(\rho\ m/\rho)-1]\ x$

The pitot tube meter is used to indicate the velocity of the fluid flow in an enclosed pipe or duct. It is very accurate and involves minimum energy losses in the flowing fluid. It requires good alignment with the flow direction to achieve best results. Pitot tube meters are able to achieve accuracy levels of better than 1% in velocity with alignment errors ofup to 15°

VENTURI METER

A venturi meter includes a cylindrical length, a converging length with an included angle of 20° or more, and short parallel throat, and a diverging section with an included angle of about 6°. The internal finishes and proportions are such to enable the most accurate readings while ensuring minimum head losses.

Assuming an inviscid fluid with no losses due to viscocity. The velocities at section 1 and 2 are u_1 and u_2. The velocities are steady and uniform over areas A_1 and A_2 The contuity exquation applies therefore $A_2\ u_2 = A_2\ u_2 = Q$

Applying bernoulli's equation to a streamline passing along the axis between the two sections(1 & 2).

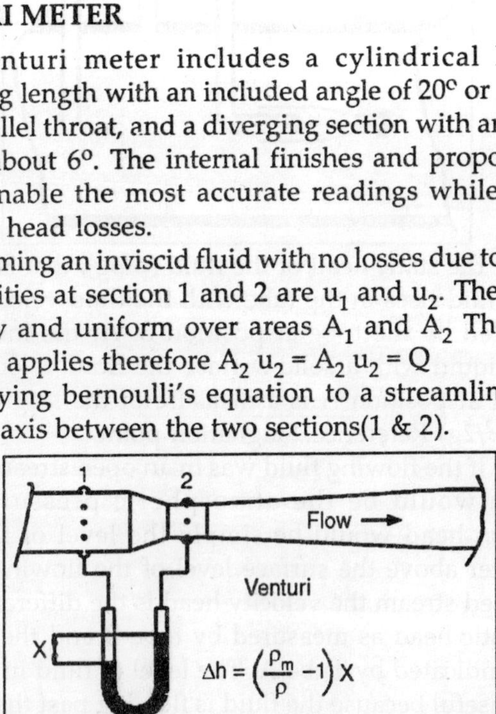

Fig. Applying Bernoulli's Equation to the Two Sections.

$$z_1 + \frac{P_1}{\rho g} + \frac{Q}{2gA} = z_2 + \frac{P_2}{\rho g} + \frac{Q}{2gA_2}$$

Therefore the ideal discharge is given by,

$$Q_{id} = \left(\frac{2g[(p_1/\rho g + z_1) - (p_2/\rho g + z_2)]}{(1/A_2^2) - (1/A_1^2)} \right)^{1/2}$$

$$= A_1 \left(\frac{2g\,\Delta h}{(A_1/A_2)^2 - 1} \right)^{1/2}$$

Note: $(p_1/\rho\, g + z_1) - (p_2/\rho\, g + z_2) = \Delta h$ is the piezometric head which can be measured by differential manometer.

Now in practice there is a slight friction loss between 1 and 2 which would result in a high Δh reading and a consequent value of Q which is too high. For real fluids therefore a factor is introduced called the coefficient of discharge factor (C_d).

For low viscosity fluids C_d = 0,98. The actual discharge as measured by a venturi is therefore given by.

$$Q_{id} = \frac{C_d A_1 A_2}{(A_1^2 - A_2^2)^{1/2}} (2g\,\Delta h)^{1/2}$$

Design and performance parameters of venturi flow meters are provided in BS EN ISO 5167-4:2003

NOZZLE FLOW METER

The nozzle as shown is practically a venturi with the diverging part removed. The basic equations used are the same as for the venturi meter.

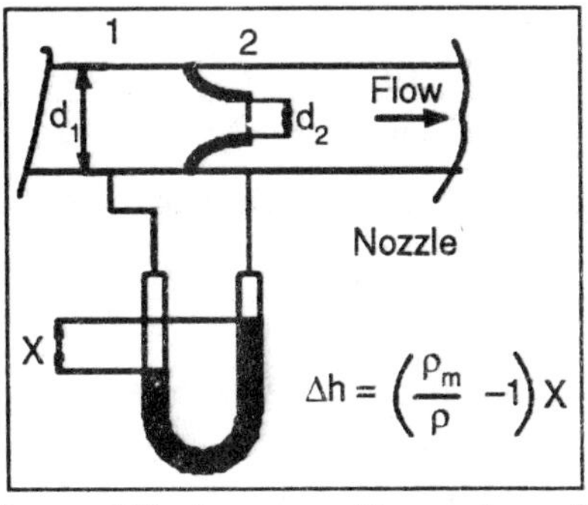

Fig. Design and Performance Parameters of Nozzles

The friction losses are slightly larger than for the venturi but this is offset by the lower cost of the unit. The fact that the manometer connections cannot be located in the ideal positions for measuring the required piezometric pressures is allowed for in selection of the coefficient of discharge factor C_d..

ORIFICE FLOW METER

The simplest and cheapest method of measuring the flow using the bernoulli equation is the sharp edged orifice as shown below.

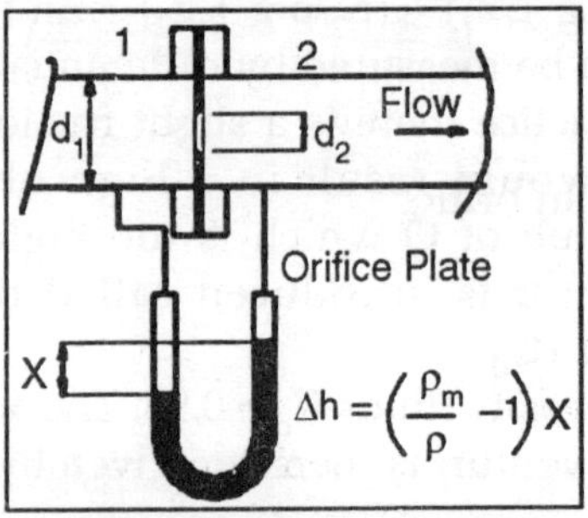

Fig. The Fluid Flow Pattern.

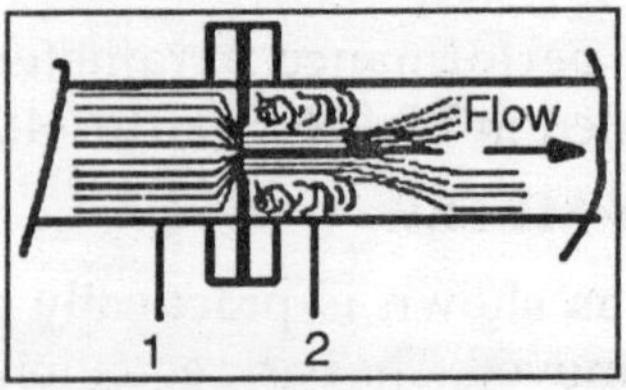

Fig. Application of Bernoulli's equation.

$$\frac{u_2^2}{2g} = \frac{p_1}{\rho g} - \frac{p_2}{\rho g} + \frac{u_1^2}{2g} = \Delta h + \frac{u_1^2}{2g}$$

Introducing for real fluid the coefficient of velocity C_v,

$$u_2 = c_v \left[2g\left(\Delta h + \frac{u_1^2}{2g}\right)\right]^{1/2}$$

Now $u_1 = Q/A_1$ and $u_2 = Q/A_c$ where A_c = The area of the vena-contracta which is the reduced area of the fluid after leaving the orifice hole. ($A_c = C_c A_2$).

A_2 = the area of the orifice and C_c is the Coefficient of contraction.

Finally C_d = Coefficient of discharge = $C_c.C_v$

Using these factors a relationship for Q can be developed from the above equation,

$$Q=\frac{C_v C_c A_2 A_1}{(A_1^2 - C_v^2 C_c^2 A_2^2)^{1/2}}(2g\Delta h)^{1/2} = \frac{C_d A_2 (2g\Delta h)^{1/2}}{[1-C_d(A_2/A_1)^2]^{1/2}}$$

To arrive at a final equation a overall discharge coefficient C is introduced.

$$C=C_d\left(\frac{1-(A_2/A_1)^2}{[1-C_d^2(A_2/A_1)^2]^{1/2}}\right)$$

Now letting $\beta = d_2 / d_1$ that is $\beta^2 = A_2 / A_1$. The equation for flow through an orifice becomes,

$$Q=\frac{C A_2(2g\Delta h)}{(1-\beta^4)^{1/2}}$$

This equation is very similar to the equation provided in BS EN ISO 5167:2 except that an expansion coefficient (ε)is introduced to cater for the measurement of compressible fluids.

The equation provided in the standards is,

$$Q_m = \frac{C}{\sqrt{(1-\beta^4)}}\varepsilon\frac{\pi}{4}d^2\sqrt{2\Delta p\rho_1} \quad ......Q_m = Q\rho_1$$

Values of the discharge coefficient C are provided in BS EN ISO 5167:2 for the different meter tapping arrangements, for different values of β against Reynold number ranges.

Small table showing C values for different Reynold numbers and β values. Detailed tables are provided in BS EN ISO 5167:2

b		**Re**	
	5×10^3	1×10^5	$1 \times 10^{8<}$
0.25	0,6102	0,6025	0,6013
0.5	0,6284	0,6082	0,6036
0.75	0,6732	0,6171	0,6025

FLOW CONDITIONING

The accuracy of the flow measuring devices is very much affected by uniformity of the approaching fluid flow. Therefore ideally there should be a straight length of piping before the flow, measuring device. It is generally accepted that for accurate flow readings there should be 50 pipe diameters of straight piping before the metering device following any pipe bend, valve, tee, reducer etc. The relevant standard provides a range of recommended minimum straight lengths depending on the nearest upstream fittings varying from 5 to 44 lengths.

This length can be reduced if flow straighteners or flow conditioning devices systems are used upstream of the flow measuring device. A flow straightener is designed to remove swirl from the flowing fluid. A flow conditioner is a device which removes swirl and also redistributes the velocity profile to produce near ideal metering conditions.

LOSSES RESULTING FROM FLOW METERING DEVICES

The orifice plate and, to a lesser extent the nozzle has significant kinetic energy losses downstream of the metering device as the locally generated kinetic energy is dissipated. The figure below illustrates the extent of these losses.

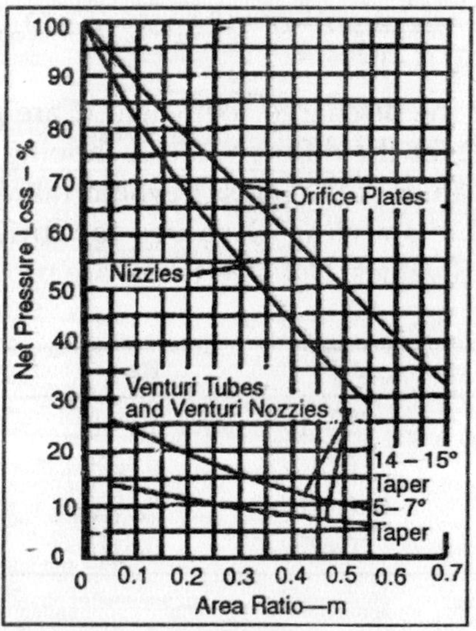

The Madden Orifice Meter gives the boiler operator precise repeatable control of the continuous blowdown flow rate. 17 different hole sizes are drilled in the stainless steel orifice plate. Each hole is precision drilled to a diameter that will produce the flow shown on he flow chart. By selecting a larger hole the continuous blowdown flow is increased, or by selecting a smaller hole number on the index band the blowdown is reduced. Four different orifice plates (1A, 2A, 3A, and 4A) are available to provide a blowdown flow range suitable for the boiler operator to maintain the target boiler water quality level.

The boiler must have a continuous blowdown skimmer pipe installed inside the boiler drum, with the opening for the pipe within 6" of the normal water level in the boiler drum. The Madden Orifice Meter may be mounted near the boiler or near a continuous blowdown heat recovery system. The constant blowdown flow provided by the Madden Orifice Meter makes it an ideal device for use in conjunction with a blowdown heat recovery system, see Madden Blowdown heat recovery system for details.

The Madden Orifice Meter has been in production for almost 50 years. The units are generally used on industrial boilers from 200 HP up to over 100,000 PPH steam production, operating at pressures from 50 psi up to 650 psi. One unit is required for each boiler. The meter can be utilized as the sole flow control device or pipe in parallel with a conductivity sensing blowdown control. The units are also used with non-fired boilers and reboilers used in refineries. The Madden Orifice Meter is guaranteed for up to 10 years against damage from cutting, wire drawing, or other distortion in the orifice plate.

Part Number	Maximum PSI	Description	Weight lbs.
OM2501A	250	Orifice meter with orifice plate 1A installed	64
OM2502A	250	Orifice meter with orifice plate 2A installed	64
OM2503A	250	Orifice meter with	64

		orifice plate 3A installed	
OM2504A	250	Orifice meter with orifice plate 4A installed	64
OM6501A	650	Orifice meter with orifice platc 1A installed	80
OM6502A	650	Orifice meter with orifice plate 2A installed	80
OM6503A	650	Orifice meter with orifice plate 3A installed	80
OM6504A	650	Orifice meter with orifice plate 4A installed	80

VENTURI METER

In comparing the venturi meter with the orifice meter, both the cost of installation and the cost of operation must be considered.

- The orifice plate can easily be changed to accomodate widely different flow rates, whereas the throat diameter of a venturi is fixed, so that its range of flow rates is circumscribed by the practical limits of Dp.
- The orifice meter has a large permanent loss of pressure because of the presence of eddies on the downstream side of the orifice-plate; the shape of the venturi meter prevents the formation os these eddies and greatly reduces the permanent loss.
- The orifice is cheap and easy to install. The venturi meter is expensive, as it must be carefully proportioned and fabricated. A home made orifice is often entirely satisfactory, whereas a venturi meter is practically always purchased from an instrument dealer.
- On the other hand, the head lost in the orifice for the same conditions as in the venturi is many times greater. The power lost is proportionally greater, and, when an orifice is inserted in a line carrying fluid continuously over long periods of time, the cost of the power may be out of all proportion to the saving

in first cost. Orifices are therefore best used for testing purposes or other cases where the power lost is not a factor, as in steam lines.

- However, in spite of considerations of power loss, orifices are widely used, partly because of their greater flexibility, because installing a new orifice plate with a different opening is a simpler matter. The venturi meter can not be so altered. Venturi meters are used only for permanent installations.
- It should be noted that for a given pipe diameter and a given diameter of orifice opening or venturi throat, the reading of the venturi meter for a given velocity is to the reading of the orifice as $(0.61/0.98)^2$, or 1:2.58.(i.e. orifice meter will show higher manometer reading for a given velocity than venturi meter).

BAND METER

Amateur Radio Bands

Just as many of us have a favorite fast food that we come back to time after time Radio Amateurs tend to have a favorite band that they use more frequently than other bands. The question of what band to use may be secondary to that favorite band. But the first question should be what band(s) am I licensed to use? From the beginning of the licensing process you will find out what bands your license covers.

Next it is a matter of having the right equipment for the bands you want to use. As an entry level license generally covers VHF(2m) and UHF(70cm) and many beginning hams will use one of both of these bands. Also hand-held single-band or dual-band radios for these bands are affordable and don't require a sophisticated antenna or power source. Most metropolitan areas also have amateur repeaters to extend the coverage when using VHF or UHF radios.

Band Restrictions

In addition to deciding what band to use there may also be restrictions within the band depending on your level of license. Canada and the U.S. each have slightly different band allocations which amateurs in each country must follow. See the sidebar for links to web sites where the band plans are defined.

Also the world is divided into 3 ITU regions each of which have their own band plan. These plans all have a great deal of overlap so you can usually talk to anyone anywhere providing you get onto a frequency acceptable to both.

Popular Bands

So what are the bands that most Radio Amateurs use? The table below shows bands that are relatively common and how and when they are in use. Again your license will determine which bands and what portions you are eligible to use.

	Band (meter)	MHz	Use*
HF	160	1.8 – 2.0	night
	80	3.5 – 4.0	night and local day
	40	7.0 – 7.3	night and local day
	30	10.1 - 10.15	CW and digital
	20	14.0 - 14.350	world wide day and night
	17	18.068 - 18.168	world wide day and night
	15	21.0 - 21.450	primarily a daytime band
	12	24.890 - 24.990	primarily a daytime band
	10	28.0 - 29.70	daytime during sunspot highs
VHF	6	50 - 54	local to world-wide
	2	144 - 148	local and medium distance
UHF	70 cm	430 - 440	local

* It should be noted that band conditions vary for many reasons and thus all of these bands can at times take on the characteristics of others. See the section on Propagation. This table should be considered a general guideline.

HF Bands

For amateurs who have an interest in long distance communication the HF (high frequency) bands will be of great interest. These bands offer propagation to all parts of the world at some time during the day, night or season. Do you want to talk to Japan from the east coast or Europe from the west? There will often be a time and a band where this communication is possible.

During a sunspot high cycle conditions are best for the higher HF frequencies and during a low the low frequencies are often in demand with all kinds of variation in between. Magazines such as QST and CQ Amateur Radio publish charts monthly that predict the best propagation to different areas of the world. These are just best guesses and will help you to choose your times and bands but there is no substitute for getting on the air.

DIMENSIONAL ANALYSIS

INTRODUCTION

Mention "dimensional analysis", "factor-label method", or "unit analysis" to students and you can sometimes hear the cries of dismay. I think this area is often terribly misunderstood by many students. It's unfortunate that many people only have their eye on the end-result rather than how they got to that point. As a teaching assistant at The Ohio State University, I'll never forget one situation with a student.

The university has a room where all the general chemistry TA's hold office hours which allows students to have access to people other than their own TA for help. One day, a student walked up and asked how to get the answer to a lab report question.

As I proceeded to work with her to help her understand how to work the question, she stopped me and said- "this is due in 15 minutes and I only want to know the answer." Well, I stopped and told her she should ask someone else because I didn't work that way. And she then walked away in a huff and probably found someone else. A case of not seeing the forest for the trees, I suppose...

Dimensional analysis is one of the most useful tools that students will learn in a science class. Not only does it encourage critical thinking, but it's something that's used every day by everybody. Yes, I mean everybody. You've already used this concept even though you may not have realized it. That quick calculation when you pulled up to the gas station with $5 in your pocket and you estimated how many gallons of gas you could get. Which laundry detergent is the best bargain? Just how many US dollars are in that Canadian dollar you're about to use in Quebec? This is not just a topic for scientific calculations.

An excellent example of just how dimensional analysis impacts our lives came in 1999 with one of NASA's Mars missions. After the loss of the Climate Observer, an investigation board was put together to determine what happened to the mission and a number of things were found to be responsible for inadequate checks and such. But, there was one factor that directly resulted in the mission's failure. I'll quote from the press release so you can read it in NASA's own words:

"The 'root' cause of the loss of the spacecraft was the failed translation of English units in a segment of ground-based, navigation-related mission software, as NASA has previously announced," said Arthus Stephenson, chairman of the Mars Climate Orbiter Mission Failure Investigation Board.

The problem was that one team used English units (feet, pounds, etc.) while another team used metric units which caused the craft to take a different path than intended. In other words, there was a problem in unit conversions- what you're learning right now!

When doing physics problems, you'll often be required to determine the numerical value and the units of a variable in an equation. The numerical value usually isn't too difficult to get, but for a novice, the same can't be said for the units.

This self-instruction unit deals with dimensional analysis, which is a useful method for determining the units of a variable in an equation. Another use of dimensional analysis is in checking the correctness of an equation which you have derived after some algebraic manipulation. Even a minor error in algebra can be detected because it will often result in an equation which is dimensionally incorrect.

The first panel lists the objectives of this unit. Read these objectives carefully.

Given the definition of a physical quantity, or an equation involving a physical quantity, you will be able to determine the dimensions and SI units of the quantity.

Given an equation, you will be able to determine if the equation is dimensionally correct or incorrect.

Most physical quantities can be expressed in terms of combinations of five basic dimensions. These are mass (M), length (L), time (T), electrical current (I), andtemperature, represented by the Greek letter theta (θ). These five dimensions have been chosen as being basic because they are easy to measure in experiments. Dimensions aren't the same as units. For example, the physical quantity, speed, may be measured in units of metres per second, miles per hour etc.; but regardless of the units used, speed is always a length divided a time, so we say that the dimensions of speed are length divided by time, or simply L/T. Similarly, the dimensions of area are L^2 since area can always be calculated as a length times a length. For example, although the area of a circle is conventionally written as πr^2, we could write it as πr_2 (which is a length) × r (another length). Now try the questions in Quiz 1.

Now that you can determine the dimensions of physical quantities, it'll be useful to write the SI units for the quantities. SI stands for International System (*Systθme Internationale*). The SI unit for mass is the kilogram, for length the metre, for time the second, for current the ampere, and for temperature the kelvin. Notice that kelvin is abbreviated as just K. The degree symbol, °, and the word "degree" are not used with kelvin.

As a quick example, let's look at speed, which has dimensions of length divided by time or L/T. Its SI units are then metres divided by seconds, represented as m/s or $m \cdot s^{-1}$. Now try Quiz 2.

Some combinations of SI units are given special names. For example, the unit of energy, $kg \cdot m^2/s^2$, is given the special name joule, which is abbreviated as J. Study the information presented below.

a) energy	joule (J)	$kg{\cdot}m^2/s^2$
b) force	newton (N)	$kg{\cdot}m/s^2$
c) frequency	hertz (Hz)	$(cycles){\cdot}s^{-1}$
d) power	watt (W)	$J/s = kg{\cdot}m^2/s^3$
e) charge	coulomb (C)	A·s

You might be wondering when to write joule, J, or $kg{\cdot}m^2/s^2$ as the energy unit. The SI convention is that if there is no number in front of the unit, then the unit is written as a full word. For example, you would write "energy is expressed in joules," with "joules" written out in full, since there is no number associated with it in the sentence. If there is a number, then "J" (or less commonly, "$kg{\cdot}m^2/s^2$") is used. Thus, you would write "the energy is 6.4 J".

Now test yourself on the SI special names. Do Quiz 3.

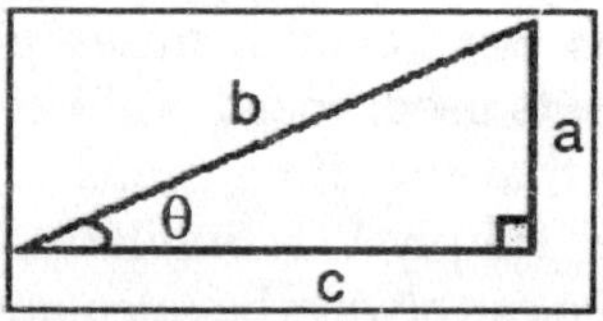

$\sin\theta = \frac{a}{b}$, so

Sin θ has Dimensions of,

$\frac{L}{L} = 1$

∴ Sin θ is Dimensionless.

Some quantities have no dimensions. For example, the sine of an angle is defined as the ratio of the lengths of two particular sides of a triangle. Thus, the dimensions of the sine are L/L, or 1. Therefore, the sine function is said to be "dimensionless". There are many other examples of "dimensionless" quantities listed in the following table.

- All trigonometric functions.
- Exponential functions.
- Logarithms.
- Angles (but notice the discussion in the next paragraph).

- Quantities which are simply counted, such as the number of people in the room plain old numbers (like 2, π, etc.).

Notice that some quantities which are "dimensionless" have units. For example, angles can be measured in units of radians or degrees, but angles are "dimensionless". Another familiar example is a frequency unit, (cycles) per second. The second, of course, is a time unit but the cycles are "dimensionless". That's the reason for cycles being written in parentheses above. Take a few moments and learn the "dimensionless" quantities above.

Try NOT to look at the table while you do it:

- We've now done all the basics so let's get into some fine points. There are two special cases of quantities which are "dimensionless". First, the argument of a trigonometric function, and second, the exponent in any exponential function. The argument of a trig function is an angle, of course, so it's "dimensionless"; and an exponent of an exponential function is the same thing as a logarithm so it's "dimensionless". These facts often are useful in helping to determine the dimensions of a quantity.

For example, if we're given that $y = e^{kt}$, where t is the time, we can state that k must have dimensions of $time^{-1}$ in order that the exponent be "dimensionless". Now let's see how well you can use this information in Quiz 5.

- A common notation, which means "the dimensions of a quantity", is simply the quantity written inside square brackets []; thus, [area] = L^2. Now you should be able to try Quiz 6.
- In an algebraic expression, all terms which are added or subtracted must have the same dimensions. This implies that each term on the left-hand side of an equation must have the same dimensions as each term on the right-hand side. For example, in the equation $a = bc + (1/2)xy$, "a" must have the same dimensions as the product "bc", and the product "(1/2)xy" must also have the same dimensions as "a" or

"bc". Remember that the "1/2" in "(1/2)xy" is just a plain old number and so it has no dimensions.

An equation in which each term has the same dimensions is said to be dimensionally correct. All equations used in any science should be dimensionally correct. The only time you'll encounter one which isn't is if there is an error in the equation. So dimensional analysis is a valuable tool in helping you to detect an equation in which you made an error in algebra, for example. Let's try this out on some equations.

$$F=-2\pi r l \frac{v}{R}\eta$$

where

F is force

r is radius

l is length

v is speed

R is distance

What are the dimensions and SI unit of h(viscosity)?

We're just about finished now. Let's do a sample question. This is a typical question which is fairly difficult. We're given an equation (shown on the left) involving force, radius, length, speed, and distance, and are asked for the dimensions and SI units of eta, (η), which is a viscosity. First, we rearrange the equation to solve for η and then convert it to an equation involving dimensions. Note that the negative sign is interpreted as "-1". Okay, you finish it off. When you've worked out the answer, move on to see if you're right or not.

$$\eta=\frac{-FR}{2\pi r l v}$$

$$[h]=\frac{[-1]\cdot[F]\cdot[R]\cdot}{[2]\cdot[\pi]\cdot[r]\cdot[l]\cdot[v]\cdot}$$

Well, the correct answer for the dimensions is $M\cdot L^{-1}\cdot T^{-1}$. The corresponding SI unit is $kg\cdot m^{-1}\cdot s^{-1}$. If you didn't get this answer, try the question again. You've

BUCKINGHAM PI THEOREM

The Buckingham Π theorem is a key theorem in

dimensional analysis.The theorem states that the functional dependence between a certain number (e.g.: n) of variables can be reduced by the number (e.g. k) of independent dimensions occurring in those variables to give a set of $p = n$ " k independent, dimensionless numbers. For the purposes of the experimenter, different systems which share the same description by dimensionless numbers are equivalent.

This theorem describes how every physically meaningful equation involving n variables can be equivalently rewritten as an equation of $n - m$ dimensionless parameters, where m is the number offundamental units used.

Most importantly, it provides a method for computing sets of dimensionless parameters from the given variables, even if the form of the equation is still unknown. However, the choice of dimensionless parameters is not unique: Buckingham's theorem only provides a way of generating sets of dimensionless parameters, and will not choose the most 'physically meaningful'.

PROVING THE Π THEOREM

Proofs of the π theorem often begin by considering the space of fundamental and derived physical units as a vector space, with the fundamental units as basis vectors, and with multiplication of physical units as the "vector addition" operation, and raising to powers as the "scalar multiplication" operation.

Making the physical units match across sets of physical equations can then be regarded as imposing linear constraints in the physical unit vector space.

The Π-theorem theorem describes how every physically meaningful equation involving n variables can be equivalently rewritten as an equation of n – m dimensionless parameters, where m is the number of fundamental units used. Furthermore, and most importantly, it provides a method for computing these dimensionless parameters from the given variables, even if the form of the equation is still unknown.

Two systems for which these parameters coincide are called similar; they are equivalent for the purposes of the equation, and the experimentalist who wants to determine the form of the equation can choose the most convenient one.

The π-theorem uses linear algebra: the space of all possible physical units can be seen as a vector space over the rational numbers if we represent a unit as the set of exponents needed for the fundamental units (with a power of zero if the particular fundamental unit is not present). Multiplication of physical units is then represented by vector addition within this vector space. The algorithm of the π-theorem is essentially a Gauss-Jordan elimination carried out in this vector space.

DIMENSIONLESS NUMBER

QUANTITY

In dimensional analysis, a dimensionless quantity (or more precisely, a quantity with the dimensions of 1) is a quantity without any physical units and thus a pure number. Such a number is typically defined as a product or ratio of quantities which do have units, in such a way that all units cancel.

EXAMPLES

"one out of every 10 apples I gather is rotten." — the rotten-to-gathered ratio is (1 apple) / (10 apples) = 0.1 = 10%, which is a dimensionless quantity. Another more typical example in physics and engineering is the measure of plane angles with the unit of "radian". An angle measured this way is the length of arc lying on a circle (with centre being the vertex of the angle) swept out by the angle to the length of the radius of the circle. The units of the ratio is length divided by length which is dimensionless.

Dimensionless quantities are widely used in the fields of mathematics, physics, and engineering but also in everyday life. Whenever one measures any physical quantity, they are measuring that physical quantity against a like dimensioned standard. Whenever one commonly measures a length with a ruler or tape measure, they are counting tick marks on the standard of length they are using, which is a dimensionless number. When they attach that dimensionless number (the number of tick marks) to the units that the standard represents, they *conceptually* are referring to a dimensionful quantity. A quantity Q is defined as the product of that dimensionless number n (the number of tick marks) and the unit U (the standard):

$$Q \overset{\text{def}}{=} nU$$

But, ultimately, people always work with dimensionless numbers in reading measuring instruments and manipulating (changing or calculating with) even dimensionful quantities.

In case of dimensionless quantities the unit U is a quotient of like dimensioned quantities that can be reduced to a number (kg/kg = 1, μg/g = 1e–6). Dimensionless quantities can also carry dimensionless units like % (= 0.01), ppm (= 1e–6), ppb (= 1e–9), ppt (= 1e–12). The CIPM Consultative Committee for Units toyed with the idea of defining the unit of 1 as the 'uno', but the idea was dropped.

PROPERTIES

- A dimensionless quantity has no physical unit associated with it. However, it is sometimes helpful to use the same units in both the numerator and denominator, such as kg/kg, to show the quantity being measured.
- A dimensionless proportion has the same value regardless of the measurement units used to calculate it. It has the same value whether it was calculated using the metric measurement system or the imperial measurement system. This doesn't hold for all dimensionless quantities; it is guaranteed to hold only for proportions.
- However, a physical quantity may be dimensionless in one system of units and not dimensionless in another system of units. For example, in the nonrationalized cgs system of units, the unit of electric charge (the statcoulomb) is defined in such a way so that the permittivity of free space $\varepsilon_0 = 1/(4\pi)$ whereas in the rationalized SI system, it is $\varepsilon_0 = 8.85419$

$\times 10^{-12}$ F/m. In systems of natural units (e.g. Planck units or atomic units), the physical units are defined in such a way that several dimensioned physical constants are made dimensionless and set to 1 (thus removing these scaling factors from equations). While this is convenient in some contexts, abolishing of all or most units and dimensions often makes practical physical calculations more error prone.

Example

The power consumption of a stirrer with a particular geometry is a function of the density and the viscosity of the fluid to be stirred, the size of the stirrer given by its diameter, and the speed of the stirrer. Therefore, we have $n = 5$ variables representing our example.

Those $n = 5$ variables are built up from $k = 3$ dimensions which are:

- *Length*: L (m)
- *Time*: T (s)
- *Mass*: M (kg)

According to the π-theorem, the $n = 5$ variables can be reduced by the $k = 3$ dimensions to form $p = n \rightarrow k = 5 \rightarrow 3 = 2$ independent dimensionless numbers which are in case of the stirrer,

- Reynolds number (This is the most important dimensionless number; it describes the fluid flow regime)
- Power number (describes the stirrer and also involves the density of the fluid)

DIMENSIONLESS QUANTITIES

There are infinitely many dimensionless quantities and they are often called numbers. Some of those that are used most often have been given names, as in the following list of examples (alphabetical order):

Name	Field of application
Abbe number	optics (dispersion in optical

	Materials)
Albedo	climatology, astronomy (reflectivity of surfaces or bodies)
Archimedes number	motion of fluids due to density differences
Bagnold number	flow of grain, sand, etc. [5]
Biot number	surface vs. volume conductivity of solids
Bodenstein number	residence-time distribution
Bond number	capillary action driven by buoyancy [6]
Brinkman number	heat transfer by conduction from the wall to a viscous fluid
Brownell Katz number	combination of capillary number and Bond number
Capillary number	fluid flow influenced by surface tension
Coefficient of static	friction of solid bodies at rest friction
Coefficient of kinetic	friction of solid bodies in translational friction motion
Courant-Friedrich-Levy	non-hydrostatic dynamics [7] number
Damkohler numbers	reaction time scales vs. transport phenomena
Darcy friction factor	fluid flow
Dean number	vortices in curved ducts
Deborah number	rheology of viscoelastic fluids
Drag coefficient	flow resistance
Eckert number	convective heat transfer
Ekman number	geophysics (frictional (viscous) forces)
Eötvös number	determination of bubble/drop shape
Euler number	hydrodynamics (pressure forces vs. inertia forces)
Fanning friction factor	fluid flow in pipes [8]
Feigenbaum constants	chaos theory (period doubling) [9]

Foppl–von Karman	thin-shell buckling number
Fourier number	heat transfer
Fresnel number	slit diffraction [10]
Froude number	wave and surface behaviour
Gain	electronics (signal output to signal input)
Galilei number	gravity-driven viscous flow
Graetz number	heat flow
Grashof number	free convection
Hagen number	forced convection
Knudsen number	continuum approximation in fluids
Kt/V	medicine
Laplace number	free convection within immiscible fluids
Lewis number	ratio of mass diffusivity and thermal diffusivity
Lockhart-Martinelli	flow of wet gases [11] parameter
Lift coefficient	lift available from an airfoil at a given angle of attack
Mach number	gas dynamics
Manning roughness	open channel flow (flow driven by coefficient gravity)
Nusselt number	heat transfer with forced convection
Ohnesorge number	atomization of liquids
Péclet number	inertial forces vs. Brownian forces
Peel number	adhesion of microstructures with substrate
Pressure coefficient	pressure experienced at a point on an airfoil
Poisson's ratio	load in transverse and longitudinal direction
Power factor	electronics (real power to apparent power)

Power number	power consumption by agitators
Prandtl number	forced and free convection
Radian	measurement of angles
Rayleigh number	buoyancy and viscous forces in free convection
Reynolds number	flow behaviour
	magnetohydrodynamics
Richardson number	effect of buoyancy on flow stability
Rockwell scale	mechanical hardness
Rossby number	inertial forces in geophysics
Schmidt number	fluid dynamics (mass transfer and diffusion) [15]
Sherwood number	mass transfer with forced convection
Sommerfeld number	boundary lubrication
Stanton number	heat transfer in forced convection
Stokes number	particle dynamics
Strouhal number	continuous and pulsating flow
van 't Hoff factor	quantitative analysis (K_f and K_b)
Weaver flame speed number	laminar burning velocity relative to hydrogen gas
Weber number	multiphase flow with strongly curved surfaces
Weissenberg number	viscoelastic flows
Womersley number	continuous and pulsating flows

DIMENSIONLESS PHYSICAL CONSTANTS

Certain physical constants, such as the speed of light in a vacuum, are normalized to 1 if the units for time, length, mass, charge, and temperature are chosen appropriately. The resulting system of units is known as Planck units. However, a handful of dimensionless physical constants cannot be eliminated in any system of units; their values must be determined experimentally.

The resulting fundamental physical constants include:

- α, the fine structure constant and the electromagnetic coupling constant
- β, the ratio of the rest mass of the proton to that of the electron
- More generally, the masses of all fundamental particles relative to that of the electron
- The strong coupling constant
- The gravitational coupling constant

REYNOLDS NUMBER

In fluid mechanics, the Reynolds number is the ratio of inertial forces ($v_s\rho$) to viscous forces (μ/L) and is used to determine whether a flow will be laminar or turbulent. It is the most important dimensionless number in fluid dynamics and provides a criterion for determining dynamic similitude. When two similar objects in perhaps different fluids with possibly different flowrates have similar fluid flow around them, they are said to be dynamically similar.

It is named after Osborne Reynolds (1842–1912), who proposed it in 1883. Typically it is given as follows for flow through a pipe:

$$Re = \frac{\rho v_s L}{\mu} = \frac{v_s L}{\nu} = \frac{\text{Intertial forces}}{\text{Viscous forces}}$$

where:

- v_s - mean fluid velocity,
- L - characteristic length (equal to diameter ($2r$) if a cross-section is circular),
- μ - (absolute) dynamic fluid viscosity,
- ν - kinematic fluid viscosity: $\nu = \mu / \rho$,
- ρ - fluid density.

Laminar flow occurs at low Reynolds numbers (Re<2100), where viscous forces are dominant, and is characterized by smooth, constant fluid motion, while turbulent flow, on the other hand, occurs at high Reynolds numbers (Re>4000) and is dominated by inertial forces, producing random eddies, vortices and other flow fluctuations.

The transition between laminar and turbulent flow is often

indicated by a critical Reynolds number (Re_{crit}), which depends on the exact flow configuration and must be determined experimentally.

Within a certain range around this point there is a region of gradual transition where the flow is neither fully laminar nor fully turbulent, and predictions of fluid behaviour can be difficult.

For example, within circular pipes the critical Reynolds number is generally accepted to be 2300, where the Reynolds number is based on the pipe diameter and the mean velocity v_s within the pipe, but engineers will avoid any pipe configuration that falls within the range of Reynolds numbers from about 2000 to 4000 to ensure that the flow is either laminar or turbulent.

For flow over a flat plate, the characteristic length is the length of the plate and the characteristic velocity is the free stream velocity.

NUSSELT NUMBER

The Nusselt number is a dimensionless number that measures the enhancement of heat transfer from a surface that occurs in a 'real' situation, compared to the heat transferred if just conduction occurred. Typically it is used to measure the enhancement of heat transfer when convection takes place.

$$Re = \frac{\rho v_s L}{\mu} = \frac{v_s L}{\nu} = \frac{\text{Intertial forces}}{\text{Viscous forces}}$$

where:

- L = Characteristic length, which is simply Volume of the body divided by the Area of the body (useful for more complex shapes)
- k_f = Thermal conductivity of the "fluid"
- h = Convection heat transfer coefficient

Selection of the significant length scale should be in the direction of growth of the boundary layer. A salient example in introductory engineering study of heat transfer would be that of a horizontal cylinder versus a vertical cylinder in natural convection.

Several empirical correlations are available that are expressed in terms of Nusselt number in the elementary analysis of flow over a flat plate etc. Sieder-Tate, Colburn and many others have provided such correlations.

For a local Nusselt number, one may evaluate the significant length scale at the point of interest. To obtain an average Nusselt number analytically one must integrate over the characteristic length. More commonly the average Nusselt number is obtained by the pertinent correlation equation, often of the form Nu = Nu(Re, Pr).

The Nusselt number can also be viewed as being a dimensionless temperature gradient at the surface.

The mass transfer analog of the Nusselt number is the Sherwood number.

PRANDTL NUMBER

The Prandtl Number is a dimensionless number approximating the ratio of momentum diffusivity and thermal diffusivity. It is named after Ludwig Prandtl.

It is defined as:

$$P_r = \frac{\nu}{\alpha} = \frac{\text{Momentum boundary – layer thickness}}{\text{Thermal boundary – layer thickness}}$$

$$= \frac{\text{viscous diffusion rate}}{\text{thermal diffusion rate}}$$

where:

- ν is the kinematic viscosity, $\nu = \mu / \rho$.
- α is the thermal diffusivity, $\alpha = k / (\rho\, c_p)$.

Typical values for Pr are:

- Around 0.7 for air and many other gases,
- Around 7 for water
- Around 7×10^{314} for earth mantle
- Between 100 and 40,000 for engine oil,
- Between 4 and 5 for R-12 refrigerant
- Around 0.015 for mercury

For mercury, heat conduction is very effective compared to convection: thermal diffusivity is dominant. For engine oil, convection is very effective in transferring energy from an area, compared to pure conduction: momentum diffusivity is dominant. In heat transfer problems, the Prandtl number controls the relative thickness of the momentum and thermal boundary layers. The mass transfer analog of the Prandtl number is the Schmidt number.

Eckert Number

The Eckert number is a dimensionless number used in flow calculations. It expresses the relationship between a flow's kinetic energy and enthalpy, and is used to characterize dissipation. It is named for the late professor Ernst R. G. Eckert.

It is defined as:

$$Ec = \frac{V^2}{c_p \Delta T}$$

where:

- V is a characteristic velocity of the flow.
- c_p is the constant-pressure specific heat of the flow.
- ΔT is a characteristic temperature difference of the flow

Stanton Number

The Stanton number is a dimensionless number which measures the ratio of heat transferred into a fluid to the thermal capacity of fluid. It is used to characterize heat transfer in forced convection flows.

$$St = \frac{h}{c_p \cdot \rho \cdot V}$$

where:

- h = Convection heat transfer coefficient.
- ρ = Density of the fluid.
- c_p = Specific heat of the fluid.
- V = Velocity of the fluid.

It can also be represented in terms of the fluid's Nusselt, Reynolds, and Prandtl numbers:

$$St = \frac{Nu}{Re \cdot Pr}$$

where:

- *Nu* is the Nusselt number.
- *Re* is the Reynolds number.
- *Pr* is the Prandtl number.

GRASHOF NUMBER

The Grashof number is a dimensionless number in fluid dynamics which approximates the ratio of the buoyancy force to the viscous force acting on a fluid. It is named after the German engineer Franz Grashof.

$$Gr \frac{g\beta(T_s - T_\infty)L^3}{\nu^2}$$

where:

g = Acceleration due to gravity.
α = Volumetric thermal expansion coefficient.
T_s = Source temperature.
T_∞ = Qquiescent temperature.
L = Characteristic length.
ν = Kinematic viscosity.

The product of the Grashof number and the Prandtl number gives the Rayleigh number, a dimensionless number that characterizes convection problems in heat transfer.There is an analogous form of the Grashof number used in cases of natural convection mass transfer problems.

$$Gr_c = \frac{g\beta^*(C_{a,s} - C_{a,a})L^3}{\nu^2}$$

where:

$$B^* = -\frac{1}{\rho}\left(\frac{\partial \rho}{\partial C_\alpha}\right)_{T,p}$$

and

g = Acceleration due to gravity
$C_{a,s}$ = Concentration of species *a* at surface
$C_{a,a}$ = Concentration of species *a* in ambient medium
L = Characteristic length
ν = Kinematic viscosity
ρ = Fluid density
C_a = Concentration of species *a*

T = Constant temperature
p = Constant pressure

SCHMIDT NUMBER

The Schmidt number is a dimensionless number approximating the ratio of momentum diffusivity (viscosity) and mass diffusivity, and is used to characterize fluid flows in where there are simultaneous momentum and mass diffusion convection processes. It was named after Ernst Schmidt.

It is defined as:

$$Sc \frac{\nu}{D}$$

where:

- ν is the kinematic viscosity
- D is the mass diffusivity.

The heat transfer analog of the Schmidt number is the Prandtl number.

BIOT NUMBER

The Biot number (Bi) is a dimensionless number used in unsteady-state (or transient) heat transfer calculations. It is named after the French physicist Jean-Baptiste Biot (1774-1862), and relates the heat transfer resistance inside and at the surface of a body.

The Biot number is defined as:

$$Bi = \frac{hL_c}{k_b}$$

where:

- h = Overall heat transfer coefficient
- L_C = Characteristic length, which is commonly defined as the volume of the body divided by the surface area of the body, such that,

$$L_c = \frac{V_{body}}{A_{surface}}$$

- k_b = Thermal conductivity of the body

The physical significance of Biot number can be fairly

understood by assuming the heat flow from a hot liquid in an cylindrical pipe (steel) to the surroundings. The heat flow experiences two resistances. One of the resistances is given by the wall of the cylindrical pipe and other by the air present near the surface of the cylindrical pipe. In this case the resistance given by air is more than the one given by the wall of the pipe. Hence Biot number is less than one. Imagine, now, that the cylindrical pipe is made of wood, which will resist the heat flow more than the air. So, here Biot number is more than one.

APPLICATIONS

Values of the Biot number larger than 0.1 imply that the heat conduction inside the body is slower than at its surface, and temperature gradients are non-negligible inside it.

MASS TRANSFER ANALOGUE

An analogous version of the Biot number (usually called the "mass transfer Biot number", or Bi_m) is also used in mass diffusion processes:

$$Bi_m = \frac{h_m L}{D_{AB}}$$

where:

- *h*: Overall mass transfer coefficient.
- L_C: Characteristic length.
- D_{AB}: Mass diffusivity.

EULER NUMBER

The Euler number or cavitation number is a dimensionless number used in flow calculations. It expresses the relationship between a flow's pressure and kinetic energy, and is used to characterize the potential of the flow to cavitate. It is named for Leonhard Euler.

It is defined as:

$$C_a = \frac{p - p_v}{\frac{1}{2}\rho V^2}$$

where:

- ρ is the density of the fluid.
- p is the local pressure.
- p_v is the vapour pressure of the fluid.
- V is a characteristic velocity of the flow.

WEBER NUMBER

The Weber number is a dimensionless number in fluid mechanics that is often useful in analysing fluid flows where there is an interface between two different fluids, especially for multiphase flows with strongly curved surfaces. It can be thought of as a measure of the relative importance of the fluid's inertia compared to its surface tension. The quantity is useful in analyzing thin film flows and the formation of droplets and bubbles. It is named after Moritz Weber (1871-1951) and may be written as:

$$\mathrm{We} = \frac{\rho v^2 l}{\sigma}$$

where:

- ρ is the density of the fluid
- v is its velocity
- l is its characteristic length
- σ is the surface tension.
- - mass diffusivity.

RAYLEIGH NUMBER

In fluid mechanics, the Rayleigh number for a fluid is a dimensionless number associated with the heat transfer within the fluid. When the Rayleigh number is below the critical value for that fluid, heat transfer is primary in the form of conduction; when it exceeds the critical value, heat transfer is primarily in the form of convection.

The Rayleigh number is named after Lord Rayleigh and is defined as the product of the Grashof number, which describes the relationship between buoyancy and viscosity within a fluid, and the Prandtl number, which describes the relationship between momentum diffusivity and thermal diffusivity.

For free convection near a vertical wall, this number is:

$$Ra_{x,c} = Gr_{r,c}\,Pr = \frac{g\beta}{\nu\alpha}(T_s - T\infty)x^3$$

where:

- Ra = Rayleigh number
- Gr = Grashof number
- Pr = Prandtl number
- g = Gravity
- x = Characteristic length
- T_s = Temperature of surface
- T_∞ = Quiescent temperature
- ν = Kinematic viscosity
- α = Thermal diffusivity
- β = Thermal expansion coefficient

For most engineering purposes, the Rayleigh number is large, somewhere around 10^6 and 10^8.

In geophysics the Rayleigh number is of fundamental importance: it indicates the presence and strength of convection within a fluid body such as the Earth's mantle, which is a solid but which behaves as a fluid over geological time scales. The low value for the Earth's mantle indicates that convection occurs throughout the mantle as a whole, and not just within mantle layers.

MACH NUMBER

Mach number (*Ma*) (pronounced: [mæk], [mQ◉k]) is a measure of relative speed. It is defined as the speed of an object relative to a fluid medium, divided by the speed of sound in that medium:

v_o/v_s

where:

v_o is the speed of the object relative to the medium and

v_s is the speed of sound in the medium

In other words, Mach number is the number of times the speed of sound an object, a duct or the fluid medium itself moves. The Mach number is named after Austrian physicist and philosopher Ernst Mach.

Overview

The Mach number is commonly used both with objects travelling at high speed in a fluid, and with high-speed fluid flows inside channels such as nozzles, diffusers or wind tunnels. As it is defined as a ratio of two speeds, it is a dimensionless number. At a temperature of 15 degrees Celsius, Mach 1 is 340.3 $m \cdot s^{''1}$ (1,225 $km \cdot h^{''1}$ or 761.2 mph) in the atmosphere. The Mach number is not a constant; it is temperature dependent. Hence in the stratosphere it remains about the same regardless of height, though the air pressure changes with height.

Since the speed of sound increases as the temperature increases, the actual speed of an object travelling at Mach 1 will depend on the fluid temperature around it. Mach number is useful because the fluid behaves in a similar way at the same Mach number. So, an aircraft travelling at Mach 1 at sea level (340.3 $m \cdot s^{''1}$, 1,225.08 km/h) will experience shock waves in much the same manner as when it is travelling at Mach 1 at 11,000 m (36,000 ft), even though it is travelling at 295 $m \cdot s^{''1}$ (654.632 MPH, 1,062 km/h, 86% of its speed at sea level).

It can be shown that the Mach number is also the ratio of inertial forces (also referred to aerodynamic forces) to elastic forces.

High-speed Flow Around Objects

High speed flight can be classified in four categories:

- *Sonic:* Ma=1
- *Subsonic:* Ma $<$ 1
- *Transonic:* 0.8 $<$ Ma $<$ 1.2
- *Supersonic:* 1.2 $<$ Ma $<$ 5
- *Hypersonic:* Ma $>$ 5

(For comparison: the required speed for low Earth orbit is ca. 7.5 $km \cdot s^{-1}$ = Ma 25.4 in air at high altitudes)

At transonic speeds, the flow field around the object includes both sub- and supersonic parts. The transonic regime begins when first zones of Ma>1 flow appear around the object. In case of an airfoil (such as an aircraft's wing), this typically happens above the wing. Supersonic flow can decelerate back to subsonic only in a normal shock; this typically happens before the trailing edge. (Fig.1a)

As the velocity increases, the zone of *Ma*>1 flow increases towards both leading and trailing edges. As *Ma*=1 is reached and passed, the normal shock reaches the trailing edge and becomes a weak oblique shock: the flow decelerates over the shock, but remains supersonic. A normal shock is created ahead of the object, and the only subsonic zone in the flow field is a small area around the object's leading edge. (Fig.1b)

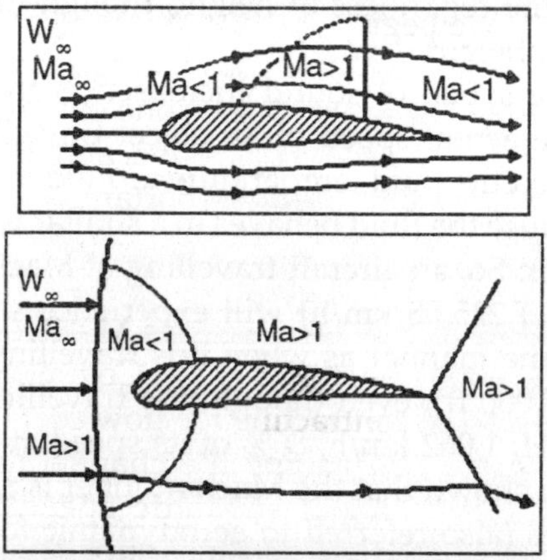

Fig. Mach number in transonic airflow around an airfoil; Ma<1 (a) and Ma>1 (b).

When an aircraft exceeds Mach 1 (i.e. the sound barrier) a large pressure difference is created just in front of the aircraft. This abrupt pressure difference, called a shock wave, spreads backward and outward from the aircraft in a cone shape (a so-called Mach cone). It is this shock wave that causes the sonic boom heard as a fast moving aircraft travels overhead. A person inside the aircraft will not hear this. The higher the speed, the more narrow the cone; at just over *Ma* =1 it is hardly a cone at all, but closer to a slightly concave plane.

At fully supersonic velocity the shock wave starts to take its cone shape, and flow is either completely supersonic, or (in case of a blunt object), only a very small subsonic flow area remains between the object's nose and the shock wave it creates ahead of itself. (In the case of a sharp object, there is no air between the nose and the shock wave: the shock wave starts from the nose.)

As the Mach number increases, so does the strength of the shock wave and the Mach cone becomes increasingly narrow. As the fluid flow crosses the shock wave, its speed is reduced and temperature, pressure, and density increase. The stronger the shock, the greater the changes. At high enough Mach numbers the temperature increases so much over the shock that ionization and dissociation of gas molecules behind the shock wave begin. Such flows are called hypersonic.

It is clear that any object travelling at hypersonic velocities will likewise be exposed to the same extreme temperatures as the gas behind the nose shock wave, and hence choice of heat-resistant materials becomes important.

High-speed Flow in a Channel

As a flow in a channel crosses M =1 becomes supersonic, one significant change takes place. Common sense would lead one to expect that contracting the flow channel would increase the flow speed (i.e. making the channel narrower results in faster air flow) and at subsonic speeds this holds true. However, once the flow becomes supersonic, the relationship of flow area and speed is reversed: expanding the channel actually increases the speed.

The obvious result is that in order to accelerate a flow to supersonic, one needs a convergent-divergent nozzle, where the converging section accelerates the flow to M = 1, sonic speeds, and the diverging section continues the acceleration. Such nozzles are called de Laval nozzles and in extreme cases they are able to reach incredible, hypersonic velocities (Mach 13 at sea level).

An aircraft Mach meter or electronic flight information system (EFIS) can display Mach number derived from stagnation pressure (pitot tube) and static pressure.

Assuming air to be an ideal gas, the formula to computer Mach number in a subsonic compressible flow is derived from the Bernoulli equation for M<1:[1]

$$M=\sqrt{5\left[\left(\frac{q_c}{p}+1\right)^{\frac{2}{7}}-1\right]}$$

where:

M is Mach number

q_c is impact pressure and

P is static pressure.

The formula to compute Mach number in a supersonic compressible flow is derived from the Rayleigh Supersonic Pitot equation:

$$M=0.88128485\sqrt{\left[\left(\frac{q_c}{p}+1\right)\left(1-\frac{1}{[7M^2]}\right)^{2.5}\right]}$$

where:

M is Mach number

q_c is impact pressure measured behind a normal shock

P is static pressure.

As can be seen, M apprears on both sides of the equation. The easiest method to solve the supersonic M calculation is to enter both the subsonic and supersonic equations into a computer spread sheet such as Microsoft Excel (or equivalent). First determine if M is indeed greater than 1.0 by calculating M from the subsonic equation. If M is greater than 1.0 at that point, then use the value of M from the subsonic equation as the initial condition in the supersonic equation. Then perform a simple iteration of the supersonic equation, each time using the last computed value of M, until M converges to a value—usually in just a few iterations.

FROUDE NUMBER

The Froude number is a dimensionless number used to quantify the resistance of an object moving through water, and compare objects of different sizes. Named after William Froude, the Froude number is based on the speed/length ratio discovered by Froude, and on which the Froude number is based.

Origins

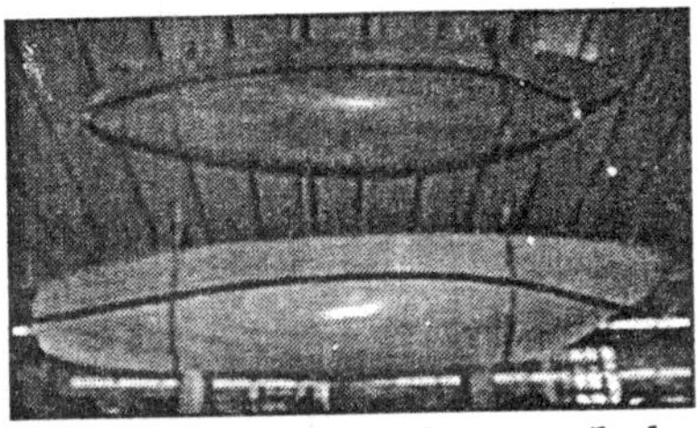

The hulls of *swan* (above) and *raven* (below). A sequence of 3, 6 and 12 (shown in the picture) foot scale models were constructed by Froude and used in towing trials to establish resistance and scaling laws.

The quantification of the resistance of floating objects is generally credited to Froude, who used a series of scale models to measure the resistance each model offered when towed at a given speed. Froude's observations led him to derive the Wave-Line Theory which first described the resistance of a shape as being a function of the waves caused by varying pressures around the hull as it moves through the water. The Naval Constructor Ferdinand Reech had put forward the concept in 1832 but had not demonstrated how it could be applied to practical problems in ship resistance. Speed/length ratio was originally defined by Froude in his *Law of Comparison* in 1868 in dimensional terms as:

$$\text{Speed lenght Ratio} = \frac{V}{\sqrt{LWL}}$$

where:

v = speed in knots

LWL is in feet

The term was converted into non-dimensional terms and was given Froude's name in recognition of the work he did. It is sometimes called Reech-Froude number after Ferdinand Reech.

DIMENSIONLESS FORMS

The dimensionless Froude number is defined as:

$$Fn = \frac{v}{\sqrt{gLWL}}$$

where v is the speed in m/s, *g* is the acceleration due to gravity, and *LWL* is the Waterline length.

The Froude number is used to compare the wave making resistance between bodies of various sizes and shapes. In fluid dynamics, the Froude number is the reciprocal of the square root of the Richardson number. When used in the context of the Boussinesq approximation it is defined as,

$$\frac{u}{\sqrt{g'h}}$$

where

g' the reduced gravity

h a representative vertical lengthscale.

Strictly, this is known as the densimetric Froude number.

Thus the Froude number is given by the ratio of inertial to gravitational forces in flow. The densimetric Froude number is usually preferred by modellers who wish to nondimensionalize a speed preference to the Richardson number which is more commonly encountered when considering stratified shear layers. For example, the leading edge of a gravity current moves with a front Froude number of about unity.

STOKES NUMBER

The Stokes number, named after Irish mathematian George Gabriel Stokes, is a dimensionless number corresponding to the behaviour of particles suspended in a fluid flow. Stokes number is defined as the ratio of the stopping distance of a particle to a characteristic dimension of the obstacle, or,

$$Stk = \frac{\tau U_o}{d_c}$$

where τ is the relaxation time of the particle, U_0 is the air velocity of the flow well away from the obstacle and d_c is the characteristic dimension of the obstacle. For Stk >> 1, particles will continue in a straight line as the gas turns around the obstacle therefore impacting on the obstacle. For Stk << 1, particles will follow the gas streamlines perfectly

FOURIER NUMBER

The Fourier number (Fo) (also known as the Fourier modulus) in physics and engineering is a dimensionless number that characterizes heat conduction. It is named after the mathematician Joseph Fourier and is defined as:

$$Fo = \frac{\alpha t}{d^2}$$

where:

- α is the thermal diffusivity [m^2/s]
- t is the characteristic time [s]
- d is the characteristic length [m]

LEWIS NUMBER

The Lewis number is a dimensionless number approximating the ratio of mass diffusivity and thermal diffusivity, and is used to characterize fluid flows in where there are simultaneous heat and mass transfer by convection.

It is defined as:

$$Le = \frac{\alpha}{D(\text{alpha over } D)}$$

(alpha over D) where α is the thermal diffusivity and D is the mass diffusivity.

The Lewis number can also be expressed in terms of the Prandtl number and the Schmidt number:

$$Le = \frac{Sc}{Pr}$$

LAPLACE NUMBER

The Laplace number (La) is a dimensionless number used in the characterisation of free surface fluid dynamics. It is related to the ratio of the surface tension to the momentum-transport inside a fluid.

It is defined as follows:

$$La = \frac{\sigma \rho L}{\mu^2}$$

where:

- σ = Surface tension
- ρ = Density
- L = Characteristic length
- μ = Absolute viscosity

FANNING FRICTION FACTOR

The fanning friction factor is a dimensionless number used in fluid flow calculations.

Defined as a friction head:

$$h_f = \frac{fv^2L}{2gR}$$

where:

- h_f is the friction loss (in head) of the pipe
- f is the fanning friction factor of the pipe
- v is the fluid velocity in the pipe
- L is the length of pipe
- g is the acceleration due to gravity
- R is the hydraulic radius of the pipe

This friction factor is one-fourth of the Darcy friction factor, so attention must be paid to note which one of these is meant in the "friction factor" chart or equation consulted. Of the two, this is the more commonly used by Chemical Engineers and those following the British convention.

PRESSURE COEFFICIENT

The pressure coefficient is a dimensionless number used in aerodynamics and fluid mechanics, most often in the design and analysis of an airfoil.

The relationship between the coefficient and the dimensional number is:

$$C_p = \frac{p - p_\infty}{\frac{1}{2}\rho V^2}$$

where:

- P∞ is the free stream pressure
- ρ is the fluid density (sea level air is 1.225kg/m^3)
- C_p of zero indicates the pressure is the same as the free stream pressure

Pressure Distribution

An airfoil at a given angle of attack will have what is called a pressure distribution. This pressure distribution is simply the pressure at 2all points around an airfoil. Typically, graphs of these distributions are drawn so that negative numbers are higher on the graph, as the C_p for the upper surface of the airfoil will usually be farther below zero and will hence be the top line on the graph.

C_l and C_p Relationship

The coefficient of lift can be calculated from the coefficient of pressure distribution by integration, or calculating the area between the lines on the distribution.

$$C_l = \int_{LE}^{TE} Cp_l(x) - C_{pu}(x)\,dx$$

where:

C_{pl} is pressure coefficient on the lower surface.

C_{pu} is pressure coefficient on the upper surface.

LE is the leading edge.

TE is the trailing edge.

When the lower surface Cp is higher(more negative) on the distribution it counts as a negative area as this will be producing down force rather than lift.

Chapter 4

Stokes's and Poiseuille's Laws

BACKGROUND

Let's begin by stating the formulas that are the goal of this analysis. They deal with the movement of viscous fluids, commonly air and water, and are some of the rare analytical results of fluid mechanics of practical value. We will study the dynamics of viscous fluids in this article, and show how these results can be obtained by analysis.

Poiseuille was interested in the flow of a fluid, actually blood, in narrow tubes, and found that the discharge Q in cc/s was related to the pressure difference Δp in dynes/cm^2 by $Q = (\pi a^4/8L\mu)\,\Delta p$. Here, a is the radius of the tube, and L its length, both in cm, and μ is the absolute viscosity of the liquid in dyne-s/cm^2 or poise.

The viscosity of water is 0.0101 poise at 20°C. Blood, with which Poiseuille was concerned, is not a simple Newtonian fluid, and its flow is a complicated problem, but water, oil and such obey the equation very well. The catch is that the flow must be rather slow, or the flow will become turbulent, obeying different rules and the resistance to flow will increase.

The flow is guaranteed to be laminar (not turbulent) provided that the Reynolds number $R = DV\rho/\mu$ is below 2000, as found from experiment. Here, D is the diameter of the tube, 2a (cm), V is the average velocity of flow $Q/\pi a^2$ (cm/s), μ the absolute viscosity, and ρ the density in gm/cc. This number is dimensionless, and is the same in any consistent system of units.

For a 1/2" pipe, R is 2000 for an average velocity of 15.7

cm/s of water, or Q = 20 cc/s, which is not a very rapid flow. We should also be aware that the laminar flow is not established immediately at the entrance to the tube, but takes a little length for this to happen, in which the liquid at the centre speeds up, and that at the wall of the pipe is retarded.

Stokes solved the problem of the steady motion of a sphere with velocity V through a still viscous fluid (or the flow of the fluid past a stationary sphere, which is the same thing). He found that the force exerted on the sphere by the fluid was $F = 6\pi\mu aV$, where F is in dyne, a in cm, V in cm/s, and μ is the absolute viscosity.

We see that the force is proportional to the velocity, the rule we often assume for a frictional force. Again, the motion must not be too fast, and a Reynolds number $R = DV\mu/\rho$ can be used to decide what is too fast.

This has the same form as in Poiseuille's equation, but the flow is different, and the critical value may be different. If V is too great, again the flow becomes turbulent, with the formation of a turbulent wake behind the sphere, and now the force is proportional to the square of the velocity.

We usually write,

$$F = C\,A\,(\rho V^2/2),$$

where,

C is the dimensionless *drag coefficient.*

Stokes's formula applies,

$$C = 24/R.$$

The critcal Reynolds number is about 1 for air, and the drag coefficient for high R is about 0.4, dropping to about 0.2 when the boundary layer changes from laminar to turbulent. This is a complex and interesting subject that you can read more about in the references.

The streamlines in Stokes flow dip in toward the sphere, so to avoid disturbing the flow pattern, there must be no nearby boundaries. A ball dropping in a tube about twice its diameter would be significantly affected (the drag would increase).

Its area of application is really to rather small spheres in air, such as cloud particles, or to slow motion in a viscous fluid, such as glycerin or oil. It was famously used by R. A. Millikan in the determination of the electronic charge, to measure the weight of the drops by how fast they fell.

The drops were so small that a correction for the mean free path of air molecules was necessary for an accurate result.

The problem can also be solved neglecting viscosity, and the results are entirely different. Some are applicable to reality, but others are not. When a sphere moves in an ideal fluid, it suffers no drag whatsoever, but once started in motion, would continue forever. The effective mass of the sphere is increased by half the mass of the fluid displaced, a result that is quite valid and observable.

It can be applied to the motion of bubbles in beer, for example. Most real-life problems of fluid flow depend greatly on viscosity, even when it would seem to have a small effect. The drag on a body moving in a fluid is mostly caused by the lower pressure in the turbulent wake behind it, and the drag is reduced by making this wake as small as possible.

The pressure on the front of the body has only a small effect. Trucks with the "streamlining" in front of the trailer would do better to put it behind!

We recall that the coefficient of viscosity μ is defined by Newton's equation F/A = i(dv/dy), where F is the shearing force on an area A, and dv/dy is the velocity gradient normal to the surface. If μ is a constant independent of the velocity of shearing, the fluid is called *Newtonian*. Many, if not most, fluids met with in engineering are Newtonian. Some, like paint and blood, are not, and there is a variety of interesting behaviors in some other substances, especially those that are a mixture of phases. μ is measured in *poise*, dyne-s/cm^2.

STRESS AND STRAIN

To begin our study, we must review the nature of stress and strain. This will require the use of quantities with a number of components in three-dimensional space, such as vectors, as well as more complex quantities. These can be represented explicitly, and were indeed for many years, but this involves tedious repetitive writing.

Vector analysis provides an excellent shorthand, and is widely understood and used. For more complex quantities, this was extended to what were called *dyadics*, but these proved rather cumbersome, and seem to have disappeared.

What turns out to be very convenient is representation by Euclidean tensors using indices. This essentially writes all the components, as was originally done, but in a very compact and useful form.

The vector a becomes the quantity a_i, where we understand that the index i takes the values 1, 2 and 3, or x, y and z, or whatever. The scalar produt of a and b is $a_i b_i$, where the repeated index implies summation over its range of values, called *contraction*.

Suppose that x_i is a point in space occupied by a solid body, and that ξ_i is the displacement of that point when the body is strained.

Then, $\varepsilon_{ij} = \partial_i \xi_j$ is the *strain*, or ratio of displacement to distance.

The symbol ∂_i is short for ∂/∂_i. This is a new kind of quantity, depending on two directions. Suppose this region of the body isn't really strained, just rotated bodily by an angle θ. Considering the displacements that result from such a rotation, it is clear that $\varepsilon_{ij} = -\varepsilon_{ji} = \theta$. We really don't want these if we are studying strains, and can get rid of them by forming $\Phi_{ij} = (\varepsilon_{ij} + \varepsilon_{ji})/2$, which is called *symmetrizing* Φ. What's left is just the *pure strain*, described by the *symmetric tensor* $\Phi_{ij} = \Phi_{ji}$.

There are six independent components of this tensor, three diagonal and three off- diagonal.

Among other things, this means that the order of the indices is immaterial, so we do not have to worry about it. In a fluid, instead of the displacements, we use the rate of displacement, the time derivative of ξ_i instead (which is just the fluid velocity v), so Φ becomes the *rate of strain* tensor.

Strain causes stress, so we must find some way to specify stress. If we imagine a plane cutting the solid, then the partA of the solid on one side of the plane exerts a force on the other side. The ratio of this force to the area considered is the *stress*.

Given a normal unit vector to specify the orientation of

the plane, then the stress is another vector that is a function of the direction of the normal vector, as well as of position.

We expect, then, that stress can be described by a quantity Ψ_{ij}. By considering the equilibrium of a small volume, the fact that it must not rotate is reflected in the *symmetry* of Ψ. Like Φ, then, Ψ is symmetric and has six independent components.

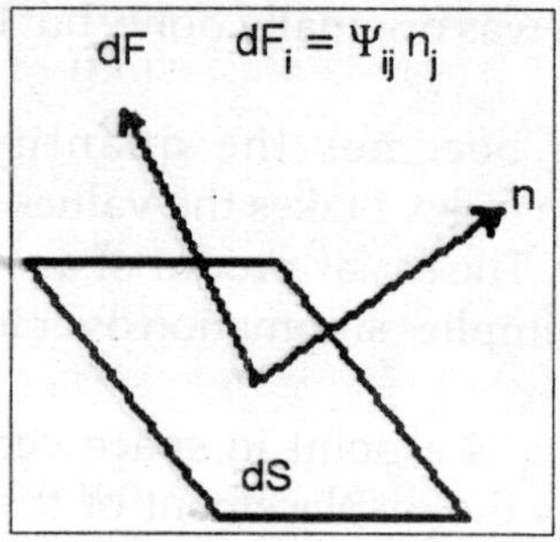

Fig. Stress Tensor

In a fluid at rest, the only stress that can exist is a pressure normal to any area, and this pressure is the same in any direction. This is, in fact, the definition of a fluid.

Therefore, considering areas normal to the coordinate axes, we find that,

$$-pn_j = \Psi_{ij}n_j$$

where n_i is a normal vector, and a pressure is considered a negative stress.

This means that $\Psi_{ij} = -p\delta_{ij}$, where δ_{ij} is the *Kronecker delta*, which is 1 when the two indices are equal, 0 otherwise. In solids or in viscous fluids, the stress may also have a component in the plane, called a *shear stress*. In this case, Ψ will acquire off-diagonal components.

In an isotropic solid, the relation between stress and strain, or between Ψ and Φ, takes the form,

$$\Psi_{ij} = [k\Phi_{kk} - (2\mu/3)\Phi_{kk}]\delta_{ij} + 2\mu\Phi_{ij}.$$

The quantity Φ_{kk} is the sum of the diagonal elements of Φ, also called its *trace*. It is equal to $\Delta V/V$, the relative volume expansion of the material, and called the *dilatation*.

The trace does not depend on the orientation of the axes, which should be clear from its meaning. In general, each of the six components of Ψ should depend on the six components of Φ, giving 36 constants of proportionality if we assume Hooke's Law, that stress is proportional to strain. If the medium is isotropic, or the same in any direction, then these reduce to only two constants, the modulus of expansion k, $p = -k(\Delta V/V)$ (positive p causes a decrease in V), and the modulus of shear, μ. Young's modulus, for example, is,

$$Y = 3k\mu/(k + \mu/3)$$

and

Poisson's ratio is,

$$\sigma = (3k - 2\mu)/(6k + 2\mu).$$

This is much simpler than it appears. Try contracting Ψ and some normal vectors and see what stresses you get, for simple Φ's.

In the case of a fluid, we find analogously that $\Psi_{ij} = [-p - (2\mu/3)\Phi_{kk}]\delta_{ij} + 2\mu\Phi_{ij}$, in terms of the pressure p and the absolute viscosity μ, which relates a shear stress to the rate of shear strain. The analogy is clear.

$$\Phi_{kk} = \mathrm{Tr}\ \Phi = ¶_k v_k = \mathrm{div}$$

v is the dilatation.

For an incompressible fluid, this quantity is zero, and the stress tensor become even simpler. Unlike solids, fluids are almost always isotropic, so all this is less of an approximation. We are now able to express the stresses in a viscous fluid in terms of the fluid velocity.

EQUATIONS OF MOTION

The motion of a fluid in the Eulerian picture is defined by the velocity field v or v_i, which is the velocity of the particle of fluid that happens to be at P(x,y,z) at time t. A time dt later, this particle has moved off a distance v_i t and has been replaced by another. If we are interested in the time rate of change of some property of the particle of fluid, which we shall call F, then dF/dt must "follow the particle," or $dF/dt = \partial F/\partial t + v_i \partial_i F$.

This is called the *substantial derivative,* and is one of the first things explained in fluid mechanics. The mass of fluid must be conserved in the motion, or $d\rho/dt + \rho\partial_i v_i = 0$, where the dilatation appears again.

Newton's Second Law for a particle of fluid is $\rho(dv_i/dt) = \rho F_i + \partial_j \Psi_{ij}$, where we have used the divergence theorem to get the force per unit volume from the stress tensor. ρF_i is an external force per unit mass, such as gravity. The total force on the particle is the sum of the external force and the force given by the stress tensor. Using the expression of Ψ in terms of Φ, and the substantial derivative, we find:

$$\rho(\text{“}v_i/\partial t) + \rho v_k \partial_k v_i = \rho F_i - \text{“}_i p + (\mu/3)\partial_i(\text{“}_k v_k) + \mu \partial_k \partial_k v_i.$$

This is called the Navier-Stokes equation, derived by Navier (1822), Poisson (1829), Saint-Venant (1843) and Stokes (1845) in successively more modern form. The second term on the left-hand side makes it nonlinear in v, which is extremely unpleasant if we are looking for analytical solutions.

If we specialize to the slow motion of an incompressible fluid, the Navier-Stokes equation becomes a bit simpler:

$$\rho(\partial v_i/\partial t) = \rho F_i - \partial_i p + \mu \partial_k \text{part;}_k v_i.$$

VORTICITY

As elements of a solid body can rotate, so can elements of a fluid. If we antisymmetrize the rate-of-strain tensor, that is form,

$$\Omega_{ij} = (\varepsilon_{ij} - \varepsilon_{ji})/2$$

we eliminate the pure strain and are left with the rotation.

This is an *antisymmetric tensor* with three independent (off-diagonal) components. These three components form a vector $(1/2)\omega = (1/2)$ curl v.

curl v is the *vorticity* of the motion of a field quantity like any other.

From Stokes's Theorem, the line integral of the velocity around a closed curve is equal to the flux of vorticity through it. The diagram shows how the antisymmetric part of the strain is related to the angle of rotation.

Taking the time derivative gives the relation between vorticity and curl v. The lines of unit length and the strains in the diagram are actually multiplied by some small number so that the change in the lengths of the axes is infinitesimal, and the strains perpendicular to both. This is hard to represent clearly in a diagram.

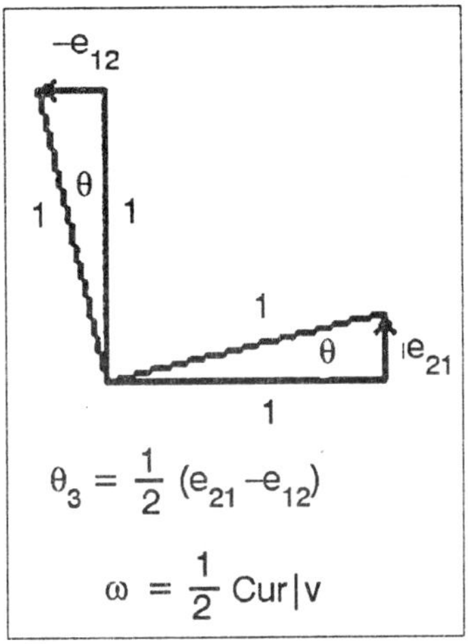

Fig. Rotation

If we take the curl of the Navier-Stokes equation (in its slow, incompressible form), we find that $\partial(\text{curl } v)/\partial t = \mu$ div grad curl v, since curl commutes with div grad (write it out, if you don't believe it).

This is the same equation satisfied by the diffusion of heat for each component of the vorticity. We see that vorticity diffuses through the fluid with the diffusion coefficient equal to the viscosity.

If the fluid is non-viscous, then $\partial(\text{curl } v)/\partial t = 0$, and curl v = constant. This is a simple form of Kelvin's Vorticity Theorem, stating that for an ideal fluid, the vorticity is a constant.

If the motion began from rest, then this says that the vorticity will be zero everywhere ever after, and the flow will be *irrotational.*

Even the smallest amount of viscosity will allow vorticity to be created, and it may dominate the flow pattern. In the present case, with Poiseuille's and Stokes's Laws, we expect vorticity.

FLOW AROUND A SPHERE

The problem is shown at the right. A rigid sphere of radius a is placed in a uniform flow of velocity V in the x-direction at large distances.

The problem is symmetrical about the x-axis, so the velocity in the xy-plane has no z component, and conditions are the same in any plane through the x-axis. We assume the flow to be steady, slow and incompressible. The velocity must satisfy the equation of continuity, div v = 0, Newton's Second Law, div grad v = (1/μ) grad p, and must vanish on the surface of the sphere. At great distances, its x-component must be V.

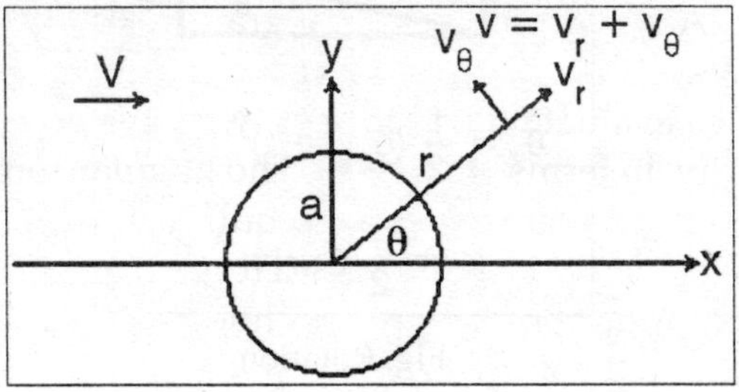

Fig. Sphere in a Uniform Flow

We'll use the vorticity, and a little luck, to solve this problem. Taking the curl of the equation of motion, we get div grad (curl v) = 0, so curl v is a solution of Laplace's equation.

From the axial symmetry of the problem, we can put curl ,

$$v = f(r,\theta)(-j\sin\varphi + k\cos\varphi),$$

where j and k are unit vectors in the y and z (out of the page) directions, and θ is the azimuthal angle around the x-axis. Then, appropriate solutions of Laplace's equation are r sin θ and sin θ/r^2, where the choice of sin θ instead of cos θ has been made so that the solution will vanish on the axis.

Substitution in Laplace's equation will verify that these are, indeed, solutions. The first solution must be thrown out because it is infinite at large distances.

Then, curl,

$$v = A(-jz/r^3 + ky/r^3) = \text{curl}\ (iA/r),$$

where z = r sin θ sin φ, y = r sin θ cos φ, and i is a unit vector in the x-direction. Therefore, we must have v = i(2A/r) + grad ψ, where ψ is some scalar function. The equation of continuity then gives div,

$v = -2A \cos\theta/r + \text{div grad}\,\psi = 0$, or $\text{div grad}\,(A\cos\theta + \psi) = 0$.

This will be satisfied provided $\psi = -A\cos\theta + \varphi$, where $\text{div grad}\,\varphi = 0$. Since φ satisfies Laplace's equation, possible solutions are $B\,r\cos\theta$ and $C\cos\theta/r^2$, where solutions proportional to $\cos\theta$ have been selected to agree with the first term in angular dependence.

Finally, then, we have,

$$v = i(2A/r) - A\,\text{grad}\cos\theta + B\,\text{grad}\,(r\cos\theta) + C\,\text{grad}\,(\cos\theta/r^2).$$

When we evaluate the gradients, we have an expression for the velocity, in terms of its radial and angular components.

This expression for v gives the required vorticity, satisfies the equation of continuity, and now if it can be made to satisfy the boundary condtions by proper choices of A, B and C, it will be the desired solution.

The behaviour at large distances requires $B = V$. The vanishing of the velocity on the surface of the sphere demands that,

$$A = -3aV/4 \text{ and } C = -a^3 V/4.$$

For the components of the velocity, we have,

$$v_r = V(1 - 3a/2r + a^3/2r^3)\cos\theta$$

$$v_\theta = -V(1 - 3a/4r - a^3/4r^3)\sin\theta.$$

The z-component of curl v in the xy-plane is $-(3aV/4r^2)\sin\theta$. Now we can use the stress tensor to find the stresses on the surface of the sphere, and by integrating them over the sphere to find the drag.

This is made much easier by choosing special x,y and z axes at a point on the sphere at angle θ with the x-axis, with the x-axis radially outward, the y axis in the direction of increasing θ, and the z-axis normal to the xy-plane, as shown in the diagram.

Then,

$$\partial v_x/\partial x = \partial v_r/r\partial\theta,\ \partial v_y/\partial y = \partial v_r/r\partial\theta + v_r/r,\ \partial v_y/\partial x = \partial v_\theta/\partial r,$$
$$\partial v_y/\partial y = \partial v_\theta/r\partial\theta - v_\theta/r.$$

Don't forget that in the second and fourth equation the unit vector also changes, and gives a contribution.

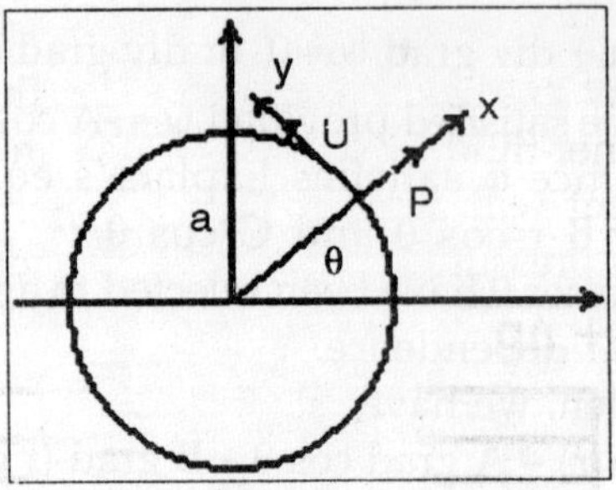

Fig. Temporary Axes

The only contributions that do not vanish on the surface of the sphere are the normal pressure, $P = (3\mu V/2a)\cos\theta$, and the shear stress,

$$U = \mu(\partial v_y/\partial x) = -(3\mu V/2a)\sin\theta.$$

The net stress on unit area is ,

$$P\cos\theta - U\sin\theta = 3\mu V/2a$$

in the x-direction, so that the total drag becomes,

$$4\pi a^2(3\mu V/2a) = 6\pi\mu aV,$$

the desired result. P and U can be separately integrated over the area, of course, and it is found that the pressure contributes 1/3, the shear 2/3 of the total drag.

Note that there is a positive pressure over the face of the sphere facing the flow, and a negative pressure over the other face. The shear force is a maximum at the equator, and is zero on the axis.

Page, and other authors, copy Lamb's pictures of the stream functions for the irrotational flow around a sphere and for the viscous flow, and remark on how very different they are, repeating Lamb.

FLOW THROUGH A TUBE

Poiseuille's problem is quite a bit easier. We can solve it using the definition of viscosity and a little statics. Consider a cylinder of fluid of thickness dr at a radius r, and of length L, and let this cylinder move at velocity v = v(r). The fluid on the side towards the wall moves a little slower, that towards the centre a little faster.

The velocity gradient causes the outer fluid to retard our cylinder, the inner fluid to urge it forward. In equilibrium, the velocity gradient will arrange itself so that the net force is balanced by the pressure difference at the ends of the cylinder.

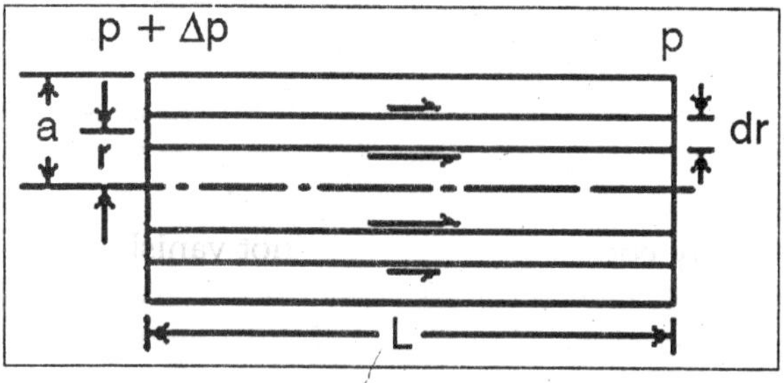

Fig. Poiseuille Flow

This is expressed mathematically by:

$(d/dr)[2\pi r L\mu(dv/dr)]dr = -2\pi r\Delta p\ dr.$

Integrating once, we find

$2\pi r L\mu(dv/dr) = -2\pi(r^2/2)\Delta p$, or $dv/dr = -(\Delta p/2L\mu)\ r + A$, where A is a constant.

By symmetry, when $r = 0$, $dv/dr = 0$, so $A = 0$.

Integrating a second time,

$v(r) = -(\Delta p/4L\mu)r^2 + B = 0,$

where B is a second constant of integration. Since $v = 0$ when $r = a$, the evaluation of B gives,

$$v(r) = (\Delta p/4L\mu)(a^2 - r^2)$$

which is the parabolic velocity profile of Poiseuille flow. As with the sphere, the vorticity forms rings around the x-axis. It is easy to find the vorticity from the velocity profile.

The rate of discharge Q is found by integrating the velocity over the cross section:

$$Q = 2\pi \int v(r)rdr.$$

The integral is easily performed after making the substitution $u = r^2$, with the result that,

$$Q = (\pi a^4/8L\mu)\Delta p.$$

The rate of discharge is proportional to the pressure difference Δp divided by the length L, or the pressure gradient, to the fourth power of the radius of the tube, and inversely proportional to the absolute viscosity. The mass discharge is proportional to the kinematic viscosity μ/ρ.

FLOW IN A PIPE

Flow in cylindrical pipes is a subject of great importance in engineering; and a great deal of empirical data has been built up that allows the solution of most practical problems with relative ease. For details, see Daugherty in the References. We will only make the connection with Poiseuille flow here.

In general, practical flow in pipes is turbulent, not laminar. The engineer speaks of "head loss" instead of pressure loss. Pressure and head are directly related through $p = \rho gh = \gamma h$, where p is pressure in dyne/cm^2, h is head in cm, ρ is the density in gm/cm^3 and γ is specific weight in dynes/cm^3. Of course, engineers use other units, such as lb/ft^2, ft, slug/ft^3 and lb/ft^3.

The head can be measured directly by a manometer at any point, though actual manometers are seldom used these days. A certain volume rate of flow Q through a length L of pipe is associated with a loss in head of h/L, a dimensionless ratio.

The engineer uses the equation,

$$h = f(L/D)(V^2/2g)$$

where D is the inside diameter of the pipe, V is the average velocity Q/A,

where,

$$A = \pi D^2/4, h$$

the head loss in a length L of pipe, and f the *Fanning friction factor*. The friction factor is a function of the Reynolds number $DV\mu/\rho$ and, for turbulent flow, of the relative roughness of the pipe, e/D, where e is the absolute roughness. The usual connection is given in the famous chart by L. F. Moody, which can be found in Daugherty.

Turbulent flow is usual when the Reynolds number exceeds 2000. This is much larger than for a sphere, because the pipe tends to keep things smooth.

Poiseuille's equation, expressed in terms of head loss, is

$$h = 32\mu LV/\gamma D^2.$$

Comparing this with the Fanning equation, we find that

$$f = 64/R,$$

where R is the Reynolds number. This is represented by a line in the Moody chart showing that f decreases steadily to about 0.032, where the critical Reynolds number is reached, and then juHmps abruptly to perhaps twice as much before it begins to decrease again as R increases.

Index